T0134832

Dual Learning

Tao Qin

Dual Learning

Tao Qin
Microsoft Research Asia (China)
Beijing, China

ISBN 978-981-15-8886-0 ISBN 978-981-15-8884-6 (eBook)
https://doi.org/10.1007/978-981-15-8884-6

This Springer imprint is published by the registered company Springer Nature Singapore Pte Ltd.
The registered company address is: 152 Beach Road, #21-01/04 Gateway East, Singapore 189721, Singapore

I would like to dedicate this book to my wife and my lovely son!

Preface

Deep neural networks have become the dominant paradigm for artificial intelligence (AI) in the past decade, and deep learning has significantly advanced various areas of AI, spanning from computer vision, natural language and speech, to game playing. A key factor contributing to the success of deep learning is availability of large-scale labeled training data. Correspondingly, a leading challenge (and also a hot research direction) for deep learning is how to learn from limited/insufficient labeled data. Dual learning is a new learning framework proposed to address this challenge by leveraging structural duality between AI tasks.

This book gives a comprehensive review of recent research on dual learning. We introduce its basic principles, including dual reconstruction, joint-probability equation, and marginal probability equation, and cover various learning settings and algorithms, including dual semi-supervised learning, dual unsupervised learning, dual supervised learning, and dual inference. For each setting, we introduce diverse applications, such as machine translation, image-to-image translation, speech synthesis and recognition, question answering and generation, image classification and generation, code summarization and generation, sentiment analysis, etc.

This book is written for researchers and graduate/undergraduate students in machine learning, computer vision, natural language, and speech areas. Background in those areas would be helpful, but it is not essential to read the book. A very brief introduction to basic machine learning and deep learning concepts is provided in Chaps. 2 and 3.

Beijing, China
July 2020

Tao Qin

Acknowledgments

This book would not have been possible without the contributions of many people.

I would like to thank my colleagues and interns at Microsoft, who have been working together with me on the topic of dual learning, including Tie-Yan Liu, Hsiao-Wuen Hon, Wei-Ying Ma, Yingce Xia, Xu Tan, Di He, Li Zhao, Fei Tian, Jiang Bian, Wei Chen, Sheng Zhao, Duyu Tang, Nan Duan, Jianxin Lin, Yijun Wang, Yi Ren, Yiren Wang, Jin Xu, and Tianyu He. I would also like to thank my external collaborators including Nenghai Yu, Zhou Zhao, Zhibo Chen, Liwei Wang, Jian Li, Enhong Chen, and Chengxiang Zhai.

I also want to thank those who allowed me to reproduce images/figures from their publications. I indicate their contributions in the figure captions throughout the text.

Acknowledgements

Contents

About the Author

Dr. Tao Qin is a Senior Principal Researcher and Manager at the Microsoft Research Asia, and an Adjunct Professor (PhD advisor) at the University of Science and Technology of China. He coined the term "dual learning" together with his colleagues and has published numerous papers on the topic in Neur IPS/ICLR/ICML/AAAI/IJCAI/CVPR/ACL/EMNLP/NAACL. His research interests include machine learning (with the focus on deep learning and reinforcement learning), artificial intelligence (with applications to language understanding, speech processing and computer vision), game theory and multi-agent systems (with applications to cloud computing, online and mobile advertising, and ecommerce), information retrieval and computational advertising.

He has published over 100 refereed papers at prestigious conferences such as Neur IPS, ICML, ICLR, AAAI, IJCAI, AAMAS, ACL, EMNLP, NAACL, KDD, WWW, SIGIR, WSDM, JAIR, EC, WINE and ACM Transactions. He served/is serving as Area Chair for AAAI, IJCAI, EMNLP, AAMAS, SIGIR and ACML, Workshop Chair for WWW 2020, and Industry Chair for DAI 2019. He is a Senior Member of the IEEE and ACM.

Chapter 1
Introduction

1.1 Motivation

Deep learning is driving and leading this wave of Artificial Intelligence (AI). Powered by deep learning, AI has made breakthroughs in different areas in recent years, including computer vision, speech recognition, natural language processing, game playing, etc.

- In 2015, ResNet [6], a very deep convolutional neural network (with 152 layers) achieved 3.57% error rate on a large scale image classification task, surpassing human error rate of 5.1%.
- In 2016, AlphaGo [13], a computer Go program based on deep neural networks and tree search, defeated a Go world champion, and became the first Go program in history that outperforms topmost human professionals.
- In 2016, a speech recognition system [24] achieved a word error rate (WER) of 5.9%, which reaches human parity and makes the same or fewer errors than professional transcriptionists, on a public conversational speech recognition dataset.
- In 2018, a translation system [4] based on deep neural networks matched human performance on a public Chinese to English new translation dataset.
- In 2019, Suphx [8], a computer Mahjong program based on deep reinforcement learning, became the first 10 DAN AI for Mahjong in history and outperformed topmost human professionals in terms of stable rank.

The success of deep learning heavily relies on massive sets of human-labeled data to train deep neural networks. As shown in Table 1.1, ResNet for image classification uses more than one million images with category labels for model

T. Qin, *Dual Learning*, https://doi.org/10.1007/978-981-15-8884-6_1

Table 1.1 Scale of human-labeled training data. We simply use DNNs (standing for deep neural networks) for those non-named systems

Task	System	Training data scale
Image classification	ResNet	Millions of images
Go playing	AlphaGo	Tens of millions of expert moves
Speech recognition	DNNs	Thousands of hours of speech
Machine translation	DNNs	Tens of millions of bilingual sentence pairs
Mahjong playing	Suphx	Tens of millions of expert actions

training, AlphaGo[1] and Suphx use tens of millions of expert moves for model training, the speech recognition system uses thousands of hours of speech with transcripts for training, and the machine translation system uses tens of millions of bilingual sentence pairs for training. Furthermore, it has been demonstrated that deep learning benefits from even more training data. Neural machine translation with tens of billions of sentence pairs can further boost the accuracy of Chinese-to-English translation over tens of millions of sentence pairs [10]. Similar conclusions have also been drawn for image classification tasks: Models trained from billions of images with category labels significantly outperform that from millions of images [9].

Unfortunately, it is usually costly to get human experts for data labeling in real-world applications. What's worse, it might be difficult to find experts for data labeling in some domains. For example, for the translation of two distant rare languages, there may be no expert in the world that can understand both rare languages. Consequently, although there might be rich labeled training data for some tasks, there are many more tasks with very limited labeled data for training. As shown in Fig. 1.1, for the translation of popular languages such as English, German, and Czech, there are tens of millions of parallel sentence pairs available for model training; in contrast, there are less than 0.2 million sentence pairs available for rare languages such as Gujarati-English translation.

Thus, how to reduce the requirement of large-scale labeled training data and how to better learn from limited labeled data are hot research areas in machine learning and especially deep learning. Different kinds of learning paradigms have been studied, including multitask learning [2, 3, 12], transfer learning [11, 17, 21], etc.

Thanks to the digitization technologies and fast development of Internet, unlabeled data is easy to collect at very low cost. Therefore, leveraging unlabeled data in machine learning and especially in deep learning is a natural solution to address the challenge of the reliance on massive sets of human-labeled training data, and has became a new trend in recent years. There are different approaches proposed to

[1] Although AlphaGo Zero [15] and AlphaZero [14] do not need expert moves for training and can learn from self-play games, those self-play games still need reward feedback, which comes from game rules and is usually unavailable in real-world applications beyond games.

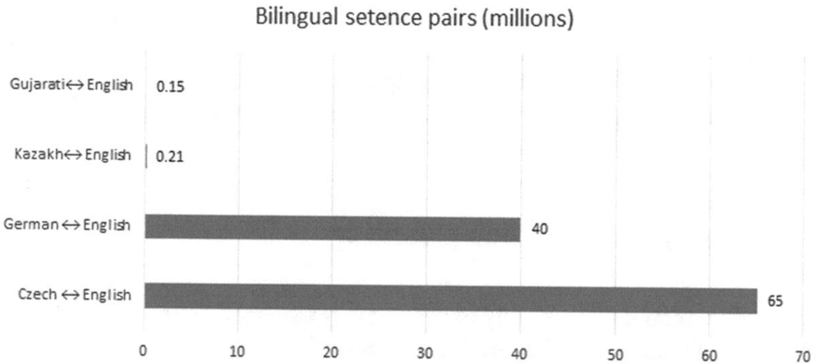

Fig. 1.1 Scale of labeled training data: numbers of bilingual sentence pairs for several language pairs in WMT 2019 [1]

leverage unlabeled data, among which dual learning [5] is a representative one and also the focus of this book.

1.2 Structure Duality in AI Tasks

Dual learning [5] is a new learning paradigm that was originally proposed to learning from unlabeled data. Later on, it was extended to broad purposes.

Definition 1.1 We say two machine learning tasks are in *dual form* if one task maps from space $\mathcal{X}$ to space $\mathcal{Y}$ and the other task maps from space $\mathcal{Y}$ to space $\mathcal{X}$. We also say the two tasks are of *structural duality*.

Many machine learning problems in real-world applications are in dual form. Here are several examples.

- Machine translation. The translation from X language (e.g., Chinese) to Y language (e.g., English) and the translation from Y language to X language are in dual form.
- Speech processing. The task of speech synthesis (text to speech) and the task of speech recognition (speech to text) are in dual form.
- Image-to-image translation. The translation of an image from X domain (e.g., photos) to Y domain (e.g., paintings) and the translation of an image from Y domain to X domain are in dual form.
- Question answering/generation. The task of question answering and the task of question generation are in dual form.
- Search and keyword advertising. The task of search, returning relevant webpages for a query/keyword, and the task of keyword advertising, suggesting relevant keywords for a webpage/website, are in dual form.

Definition 1.2 If two tasks are in dual form, we call the task mapping space $\mathcal{X}$ to space $\mathcal{Y}$ the *primal task* or *forward task*, correspondingly its model the *primal model* or *forward model*, and the task mapping space $\mathcal{Y}$ to space $\mathcal{X}$ the *dual task* or *backward task*, correspondingly its model the *dual model* or *backward model*.

1.3 Categorization of Dual Learning

Although structural duality widely exists in many real-world applications, it has not been well explored and studied until recent years [5].

Roughly speaking, the basic idea of dual learning is to leverage the symmetric (primal-dual) structure of machine learning tasks to obtain effective feedback or regularization signals to enhance the learning or inference process.

The research on dual learning can be categorized from different perspectives.

1.3.1 Classified by Data Settings

According to the data used for model training, the work on dual learning can be classified into following categories.

- *Dual semi-supervised learning* [5, 18–20], in which both labeled data and unlabeled data are used for dual training.
- *Dual unsupervised learning* [7, 25, 26], in which only unlabeled data is used for dual training.
- *Dual supervised learning* [16, 23], in which only labeled data is used for dual training.

Different from those learning settings, in which structural duality is utilized in training, *dual inference* [22] leverages structure duality in inference.

1.3.2 Classified by Principles

Depending on application scenarios, different principles have been proposed and studied to leverage structural duality.

1. Dual reconstruction principle. Intuitively, given a data sample x in space $\mathcal{X}$, after first applying the forward model f and then the backward model g, we should be able to reconstruct the original sample x. This principle can be implemented deterministically [7, 25, 26]:

$$x = g(f(x)), \tag{1.1}$$

or probabilistically [5, 18]:

$$x = \arg\max_{x'} P(x'|g(f(x))). \tag{1.2}$$

2. Joint-probability [22, 23] and marginal-probability principle [19, 20]. Consider the joint probability of a data pair (x, y) where $x \in \mathcal{X}$ and $y \in \mathcal{Y}$. Let $P(x)/P(y)$ indicate the marginal distribution of x/y respectively, $P(y|x; f)$ indicate the conditional probability of generating y from x using the forward model f, and $P(x|y; g)$ indicate the conditional probability of generating x from y using the backward model g. Intuitively, we should have

$$P(x, y) = P(x)P(y|x; f) = P(y)P(x|y; g). \tag{1.3}$$

We leave the marginal-probability principle to Chap. 9.

1.4 Book Overview

This book can be useful for a variety of readers, and we specifically target at two kinds of audiences. One kind of these target audiences is university (undergraduate or graduate) students who are learning or doing research on machine learning, natural language processing, and computer vision. The other kind is industry practitioners such as AI engineers and data scientists who are working on AI related products. To best accommodate audiences with diverse background, we organize this book into five parts.

Part I introduces basic concepts of machine learning (Chap. 2) and deep learning (Chap. 3). Readers with machine learning and deep learning background should feel free to skip this part.

Part II focuses on dual learning based on the principle of dual reconstruction. Given that dual learning with the reconstruction principle has been studied in diverse applications, we organize the chapters in this part based on applications so that readers can selectively read chapters relevant to their interests and background. The applications covered in this part include machine translation and other natural language tasks (Chap. 4), image translation and other computer vision tasks (Chap. 5), and speech processing (Chap. 6).

Part III focuses on dual learning based on the probability principle. Chapter 7 introduces the joint-probability principle for dual supervised learning, Chap. 8 describes dual inference based on the joint-probability principle, and Chap. 9 covers the marginal-probability principle for semi-supervised learning.

After introducing the algorithms and applications, we turn to several advanced topics in Part IV. Chapter 10 focuses on theoretical understandings of the dual reconstruction principle. Chapter 11 discusses the connections between dual learning and several other learning paradigms.

Part V summarizes this book and discusses future research directions (Chap. 12).

References

1. Barrault, L., Bojar, O., Costa-Jussà, M.R., Federmann, C., Fishel, M., Graham, Y., et al. (2019). Findings of the 2019 conference on machine translation (WMT19). In *Proceedings of the Fourth Conference on Machine Translation (Volume 2: Shared Task Papers, Day 1)*, Florence (pp. 1–61). Stroudsburg: Association for Computational Linguistics.
2. Caruana, R. (1997). Multitask learning. *Machine Learning, 28*(1), 41–75.
3. Evgeniou, T., & Pontil, M. (2004). Regularized multi-task learning. In *Proceedings of the Tenth ACM SIGKDD International Conference on Knowledge Discovery and Data Mining* (pp. 109–117).
4. Hassan, H., Aue, A., Chen, C., Chowdhary, V., Clark, J., Federmann, C., et al. (2018). Achieving human parity on automatic Chinese to English news translation. arXiv:1803.05567.
5. He, D., Xia, Y., Qin, T., Wang, L., Yu, N., Liu, T.-Y., et al. (2016). Dual learning for machine translation. In *Advances in neural information processing systems* (pp. 820–828).
6. He, K., Zhang, X., Ren, S., & Sun, J. (2016). Deep residual learning for image recognition. In *Proceedings of the IEEE Conference on Computer Vision and Pattern Recognition* (pp. 770–778).
7. Kim, T., Cha, M., Kim, H., Lee, J. K., & Kim, J. (2017). Learning to discover cross-domain relations with generative adversarial networks. In *Proceedings of the 34th International Conference on Machine Learning* (Vol. 70, pp. 1857–1865). JMLR.org
8. Li, J., Koyamada, S., Ye, Q., Liu, G., Wang, C., Yang, R., et al. (2020). Suphx: Mastering Mahjong with deep reinforcement learning. Preprint. arXiv:2003.13590.
9. Mahajan, D., Girshick, R., Ramanathan, V., He, K., Paluri, M., Li, Y., et al. (2018). Exploring the limits of weakly supervised pretraining. In *Proceedings of the European Conference on Computer Vision (ECCV)* (pp. 181–196).
10. Meng, Y., Ren, X., Sun, Z., Li, X., Yuan, A., Wu, F., et al. (2019). Large-scale pretraining for neural machine translation with tens of billions of sentence pairs. arXiv:1909.11861.
11. Pan, S. J., & Yang, Q. (2010). A survey on transfer learning. *IEEE Transactions on Knowledge and Data Engineering, 22*(10), 1345–1359.
12. Ruder, S. (2017). An overview of multi-task learning in deep neural networks. Preprint. arXiv:1706.05098.
13. Silver, D., Huang, A., Maddison, C. J., Guez, A., Sifre, L., Van Den Driessche, G., et al. (2016). Mastering the game of go with deep neural networks and tree search. *Nature, 529*(7587), 484.
14. Silver, D., Hubert, T., Schrittwieser, J., Antonoglou, I., Lai, M., Guez, A., et al. (2018). A general reinforcement learning algorithm that masters chess, shogi, and Go through self-play. *Science, 362*(6419), 1140–1144.
15. Silver, D., Schrittwieser, J., Simonyan, K., Antonoglou, I., Huang, A., Guez, A., et al. (2017). Mastering the game of go without human knowledge. *Nature, 550*(7676), 354–359.
16. Sun, Y., Tang, D., Duan, N., Qin, T., Liu, S., Yan, Z., et al. (2019). Joint learning of question answering and question generation. *IEEE Transactions on Knowledge and Data Engineering, 32*(5), 971–982.
17. Torrey, L., & Shavlik, J. (2010). Transfer learning. In *Handbook of research on machine learning applications and trends: algorithms, methods, and techniques* (pp. 242–264). Pennsylvania: IGI Global.
18. Wang, Y., Xia, Y., He, T., Tian, F., Qin, T., Zhai, C. X., et al. (2019). Multi-agent dual learning. In *Seventh International Conference on Learning Representations, ICLR 2019*.
19. Wang, Y., Xia, Y., Zhao, L., Bian, J., Qin, T., Liu, G., et al. (2018). Dual transfer learning for neural machine translation with marginal distribution regularization. In *Thirty-Second AAAI Conference on Artificial Intelligence*.
20. Wang, Y., Xia, Y., Zhao, L., Bian, J., Qin, T., Chen, E., et al. (2019). Semi-supervised neural machine translation via marginal distribution estimation. *IEEE/ACM Transactions on Audio, Speech, and Language Processing, 27*(10), 1564–1576.

21. Weiss, K., Khoshgoftaar, T. M., & Wang, D. D. (2016). A survey of transfer learning. *Journal of Big Data, 3*(1), 9.
22. Xia, Y., Bian, J., Qin, T., Yu, N., & Liu, T.-Y. (2017). Dual inference for machine learning. In *Proceedings of the 26th International Joint Conference on Artificial Intelligence* (pp. 3112–3118).
23. Xia, Y., Qin, T., Chen, W., Bian, J., Yu, N., & Liu, T.-Y. (2017). Dual supervised learning. In *Proceedings of the 34th International Conference on Machine Learning* (Vol. 70, pp. 3789–3798). JMLR.org
24. Xiong, W., Droppo, J., Huang, X., Seide, F., Seltzer, M., Stolcke, A., et al. (2016). Achieving human parity in conversational speech recognition. Preprint. arXiv:1610.05256.
25. Yi, Z., Zhang, H., Tan, P., & Gong, M. (2017). Dualgan: Unsupervised dual learning for image-to-image translation. In *Proceedings of the IEEE International Conference on Computer Vision* (pp. 2849–2857).
26. Zhu, J.-Y., Park, T., Isola, P., & Efros, A. A. (2017). Unpaired image-to-image translation using cycle-consistent adversarial networks. In *Proceedings of the IEEE International Conference on Computer Vision* (pp. 2223–2232).

Part I
Preparations

Dual learning is a sub branch of machine learning. To well understand dual learning, one needs to have some basic knowledge on machine learning and understand the basic concepts of machine learning. Furthermore, most works on dual learning are based on deep neural networks, and therefore good background knowledge on deep learning will be very helpful to understand dual learning. Thus, in the first part of this book, we give a brief introduction to machine learning and deep learning.

Chapter 2
Machine Learning Basics

This chapter introduces several basic concepts of machine learning that are relevant to dual learning and will be used throughout the rest of this book. Readers who are new to machine learning and want a more comprehensive coverage are encouraged to check machine learning textbooks such as [2, 13, 29].

2.1 Machine Learning Paradigms

Roughly speaking, an algorithm is a computer program that can process an input and produce an output. Mathematically, an algorithm is a a function

$$f : \mathcal{X} \rightarrow \mathcal{Y}$$

that maps from the input space $\mathcal{X}$ to the output space $\mathcal{Y}$. For example, a sorting algorithm is an algorithm that takes a set of elements as input and outputs an ordered list with the same set of elements.

A machine learning algorithm is a special kind of computer programs that can learn from data and output an algorithm/function. In other words, a machine learning algorithm takes data as input and outputs a function $f : \mathcal{X} \rightarrow \mathcal{Y}$.[1] In most cases, we call the output of a machine learning algorithm the *model*, which can be applied to unseen data.

Depending on the strength of feedback signal in data, machine learning can be categorized into three major paradigms: supervised learning, unsupervised learning, and reinforcement learning.

[1]A more formal definition of machine learning algorithms is provided in [28]: "A computer program is said to learn from experience E with respect to some class of tasks T and performance measure P if its performance at tasks in T, as measured by P, improves with experience E."

T. Qin, *Dual Learning*, https://doi.org/10.1007/978-981-15-8884-6_2

2.1.1 Supervised Learning

In supervised learning, the data is a set of n pairs $\{(x_1, y_1), (x_2, y_2), \cdots, (x_n, y_n)\}$, where $x_i \in \mathcal{X}$ and $y_i \in \mathcal{Y}$, and a learning algorithm learns a function $f()$ from the data mapping from the input space $\mathcal{X}$ to the output space $\mathcal{Y}$. The learning process is called training, the data set $\{x_i, y_i\}_{i=1}^n$ is called training dataset, and a pair $\{x_i, y_i\}$ in the training set is a training example/sample, where x_i is the input (usually a feature vector) of the example, and y_i is the desired output (or label) of the example. After training, the learnt function is applied to unseen input $x_j \in \mathcal{X}$ to make predictions.

Supervised learning is the most widely studied learning paradigm and plays an important role in many real-world applications.

- *Spam email detection* is a popular functionality in almost every email service, in which one designs an automatic spam detector that can filter spam emails before getting into users' inboxes. Here x_i is the features extracted from an email, e.g., whether the email contains a specific word, the domain of the sender, etc.; y_i is 1 indicating a spam email or 0 indicating non-spam email.
- *Image classification* has many practical applications in our daily life. (1) Personal photo organization. Nowadays many photos are taken and stored in every smart phone. Classifying and organizing photos into several categories (e.g., whether a photo is indoor or outdoor, contains flowers, etc.) greatly ease the access of those photos. (2) Optical character recognition, which identifies different characters from images of handwritten, typewritten or printed text, plays a key role in the digitization of physical documents. (3) Face recognition, which is to identify a person from an image or video frame, is used as access control in smart phones and computers, and used in security systems such as video surveillance.
- *Machine translation*, which translates one language to another language by computer/algorithms, is widely used in cross-lingual communications, including international conferences, cross-country business meetings, and cross-lingual search/reading/subtitling.
- *Automatic speech recognition*, which automatically recognizes and translates spoken language into text by computer/algorithms, is a basic component in many human-computer interaction systems to understand human language and interact with human beings. Those systems include mobile and home virtual assistants (e.g., Amazon Alexa, Apple Siri, Google Assistant, Microsoft Cortana), in-car systems, meeting transcription systems, and subtitling for films, TV programs and video games.
- *Text to speech (speech synthesis)*, the reverse task of automatic speech recognition that converts text to speech, is a basic component in the above listed human-computer systems for them to respond and interact with human beings.

There are several typical supervised learning problems, depending on the format of $\mathcal{Y}$.

- *Classification:* In classification problems, the output space is a set of k categories: $\mathcal{Y} = \{1, 2, \ldots, k\}$. The learning algorithm needs to produce a function that

can identify one or multiple of the k categories an input belongs to: $f : \mathcal{X} \rightarrow \{1, 2, \ldots, k\}$. The input $x \in \mathcal{X}$ can be of different formats, depending on applications. For example, in traditional classification problems [38], x is usually a d-dimensional vector and $\mathcal{X} = \mathcal{R}^d$; in image classification [23], x could be a tensor (e.g., 3D matrix) representation of an image; in sequence classification [52], x could be a sequence of words and of variable length.
- *Regression:* In regression problems, given an input x, the function f needs to predict a continuous variable y, i.e., $\mathcal{Y} = \mathcal{R}$. Linear regression, where f is a linear function of x, has been well studied in both Statistics [39, 51] and machine learning [13]. Again, the format of input x depends on applications. In most regression problems, the input x is a feature vector with fixed dimensions, i.e., $\mathcal{X} = \mathcal{R}^d$; in some image related applications such as predicting the age of a face image,[2] x is the tensor (e.g., 3D matrix) representation of an image.
- *Structure prediction:* In many problems, y is more complex than a simple category label or a real value: it can be a sequence of words (such as in machine translation [1, 22] and text summarizing [15, 30]), and a ranking list of objects (such as in information retrieval [5, 26] and recommender systems). If y is a sequence, the problem is usually called *sequence generation*; if both x and y are sequences, the problem is called *sequence to sequence (seq2seq)* learning.

Dual learning has been studied in both supervised classification problems (e.g., image classification, sentiment classification) and supervised sequence to sequence learning problems (e.g., machine translation, speech recognition and synthesis, question answering and generation, code summarization and generation), which will be introduced in Chap. 7.

2.1.2 Unsupervised Learning

In unsupervised learning, data are given with x but no label y, and a machine learning algorithm is used to automatically find patterns from data. Here are several typical unsupervised learning problems.

- *Clustering:* Roughly speaking, clustering [17, 18] is to group data points. Given a set of data points, a clustering algorithm classify/cluster them into several different groups, so that the data points in the same group should be similar to each other and those in different groups should be dissimilar to each other. Clustering is a common technique for unsupervised data analysis and has been used in many fields, such as document clustering that group documents/webpages of similar topics [19, 31, 53], gene clustering that groups genes with similar expression levels [24, 46], image clustering that groups images of similar low-

[2]https://www.how-old.net.

level features [34] or high-level semantics [4, 14], and climate data analysis that is used to discover climate indices [44].

- *Dimension reduction* [3, 11] refers to representing data with less features or lower dimensions. In real-world applications, data is usually of high dimensionality. Dimension reduction can help human to better understand, analysis and visualize data.
- *Self-supervised learning:* Different from clustering and dimension reduction that have been well studied in classical machine learning, self-supervised learning is a relatively new concept proposed in recent years. Rather than requiring external labels provided by human in supervised learning, self-supervised learning trains a model using labels that are naturally embedded in the input data. Self-supervised learning has become a popular approach in computer vision tasks [10, 40, 47] and natural language processing tasks [8, 33]. The key of self-supervised learning is to define the implicit labels embedded in the input data, which consequently defines the self-supervised learning task.

A large part of dual learning studies is conducted in the setting of unsupervised learning, which will be introduced in Chaps. 4 and 5.

2.1.3 Reinforcement Learning

Although dual learning is not directly related to reinforcement learning, some dual learning algorithms [16, 36] leverage reinforcement learning algorithms for model training. Therefore, we briefly introduce reinforcement learning in this sub section.

Situated in between supervised learning and unsupervised learning, reinforcement learning [45] deals with sequential decision making problems with limited and usually delayed feedback. Different from supervised learning, in which a learning algorithm is told the correct label y for an input x, in reinforcement learning, a learning agent facing a state s is not told which is the correct action; instead it needs to take a sequence of actions, interacts with the environment, and adjust its behavior/policy according to the delayed reward signal provided by the environment. The goal of a reinforcement learning algorithm is to optimize its policy so as to maximize accumulated reward.

Reinforcement learning has a long and rich history that can be dated back to the research on trial-and-error learning started in the psychology of animal learning and the research on optimal control using value functions and dynamic programming. Enhanced by deep learning, deep reinforcement learning has received great attention in recent years because of its huge achievement in game playing such as AlphaGo [41], AlphaGo Zero [43], AlphaZero [42], and Suphx [25]. There are multiple types of deep reinforcement learning algorithms such as model-free algorithms, model based algorithms, value based algorithms, and policy optimization. Of them, policy optimization methods are often used in other machine learning problems such as supervised learning when the objective function is not continuous with respect to model parameters, which is also the case for dual learning [16, 36].

2.1.4 More Learning Paradigms

In addition to supervised learning, unsupervised learning, and reinforcement learning, there are also other machine learning paradigms. Here we briefly mention some of them.

Since labeled data is usually costly to obtain and unlabeled data is of low low and almost unlimited, unlabeled data is often used to enhance the learning from labeled data. In between supervised learning and unsupervised learning, *semi-supervised learning* [55] uses both labeled data and unlabeled data for training.

Transfer learning [32] aims to boost the performance of a target problem, which is usually of limited training data, by transferring knowledge from a related problem or domain, which has rich labeled data.

Multitask learning [6] trains models for multiple learning tasks at the same time, exploiting commonalities and differences across tasks, so as to improve learning efficiency and prediction accuracy for individual models.

As will be seen in Chap. 11, dual learning is related to but different from semi-supervised learning, transfer learning, and multitask learning. More importantly, it can be combines with those learning paradigms, leading to dual semi-supervised learning [16] and dual transfer learning [50], which will be covered in later chapters.

2.2 Key Components of a Learning Algorithm

As supervised learning is most widely studied in machine learning, we take supervised learning as example to present the key components of a learning algorithm. A simplified descriptive definition of a supervised learning algorithm is as follows:

Definition 2.1 A supervised learning algorithm is a computer program that aims to find a function f from a function space $\mathcal{F}$ that minimizes a loss function $l()$ over a training dataset D:

$$\min_{f \in \mathcal{F}} \frac{1}{|D|} \sum_{(x,y) \in D} l(f(x), y), \tag{2.1}$$

where $|D|$ denotes the number of data points in the training dataset D, $\mathcal{F}$ is the function space containing functions mapping from the input space $\mathcal{X}$ to the output space $\mathcal{Y}$, and $l()$ is the loss function that evaluate the difference between the prediction of function f and the label y.

As shown in above definition, a supervised learning algorithm consists of several typical components: the training dataset D, a hypothesis space $\mathcal{F}$, a loss function $l()$, and an optimizer to perform the min operation.

In supervised learning, training data points are pre-given and in the format of input-output pairs. That is, the training dataset is usually a set of n pairs: $D = \{(x_i, y_i)\}_{i=1}^n$. As aforementioned, the input and output can be of different format (e.g., vectors, matrices/tensors, sequences, etc.) in different applications.

In early years of machine learning (before deep learning), feature engineering is very important while preparing training data. For the example of image classification, classical machine learning algorithms do not directly take images with raw pixels as inputs. Instead, researchers have designed different kinds of feature extractors (e.g., the local scale-invariant features [27]) to represent an image as a feature vector.

The hypothesis space $\mathcal{F}$ determines what kind of functions we want to learn from the data. Typical examples include linear functions, decision trees, shallow or deep neural networks. A function in the hypothesis is also called a predictor function or a model. Generally speaking, the hypothesis should be complex enough to well express and fit the data but not too complex to overfit the data. Dual learning is most based on deep neural networks, which will introduced in Chap. 3.

The loss function $l()$ takes the prediction $f(x)$ of a function and the ground truth label y of the data x as inputs and outputs a real value indicating the goodness of the prediction $f(x)$. Typical choices of the loss function include the square error loss in regression problems, the hinge loss in support vector machines [37], the exponential loss in gradient boosting trees [12], and the negative log-likelihood loss in classification problems with probabilistic classifiers.

Different optimizers have been designed to solve Eq. (2.1) for different loss functions and hypothesis spaces, especially when learning from large-scale training data. For example, sequential minimal optimization [35] is proposed for fast training of support vector machines, XGBoost [7] and LightGBM [20] are designed for scalable training of gradient boosting decision trees, and many variants of stochastic gradient descent including AdaGrad [9], AdaDelta [54] and Adam [21] are proposed to train deep neural network models.

In the following, we will take multiclass logistic regression,[3] a classification algorithm that generalizes logistic regression for two-class classification to multi-class problems, as example to illustrate the above components.

[3]Multiclass logistic regression has been studied under different names, including multinomial logistic regression, softmax regression, polytomous logistic regression, etc.

2.2.1 An Example: Multiclass Logistic Regression

Let's consider a K-class classification problem, i.e., $\mathcal{Y} = \{1, 2, \cdots, K\}$. Suppose the input is a d-dimensional feature vector, i.e., $\mathcal{X} = \mathcal{R}^d$.

Multiclass logistic regression constructs a linear function f_k that assigns a score $f_k(x)$ for each class k:

$$f_k(x; W) = w_k \cdot x, \tag{2.2}$$

where $w_k \in \mathcal{R}^d$ is a weight vector associated with the class k. Let $W \in \mathcal{R}^{d\times K}$ is a matrix whose k-th column is w_k. Then the prediction function of multinomial logistic regression can be written as

$$f(x; W) = W^T x, . \tag{2.3}$$

where $f(x; W)$ is a K-dimensional vector and its k-th dimension $f_k(x)$ is the score for the k-th class.

The predicted outcome/class for x is the class with the highest score. In other words, the larger the score $f_k(x; W)$ is, the more likely x belongs to the class k. To formally describe the correlation between the score and the likelihood, we use the softmax function to convert the scores to probabilities over K classes:

$$p_k(x; f) = \frac{\exp(f_k(x; W))}{\sum_{j=1}^{K} \exp(f_j(x; W))} = \frac{\exp(w_k \cdot x)}{\sum_{j=1}^{K} \exp(w_j \cdot x)}. \tag{2.4}$$

The training process of multinomial logistic regression is to maximize the log likelihood of the correct class label y_i:

$$\log p_{y_i}(x; f) = \log \frac{\exp(f_{y_i}(x; W))}{\sum_{j=1}^{K} \exp(f_j(x; W))}. \tag{2.5}$$

Therefore, the loss function in multi-class logistic regression is

$$l(y_i, f(x; W)) = -\log p_{y_i}(x; f), \tag{2.6}$$

which is called negative log likelihood loss. The overall loss over the training dataset is

$$L(f) = \frac{1}{|\mathcal{D}|} \sum_{(x,y)\in\mathcal{D}} l(y, f(x; W)). \tag{2.7}$$

When the size of training data is not very large, gradient descent is often used to optimize the above loss and learn the parameters W. Gradient descent is a first-order iterative optimization method, which is known to be able to find a local minimum of a differentiable function if the there exists one. We randomly initialize the parameter matrix W_0 and then iteratively update it:

$$W_{t+1} = W_t - \gamma_t \frac{\partial L(f)}{\partial W_t}, \tag{2.8}$$

where $\gamma_t \in \mathbb{R}_+$ is the step size at step t. The iteration terminates when certain criterion is met, e.g., the maximal iteration number is achieved, or the loss on a held-out validation set does not decrease any more.

2.3 Generalization and Regularization

A key challenge in machine learning is that a machine learning algorithm needs to perform well on unseen data, not just on the data used for model training. This ability is referred to as generalization, i.e., generalizing well to "out of sample data" while trained from "in sample data". In machine learning, the error measure on the training data is called training error, and the error on unseen data is called test error, or generalization error, which is usually estimated on a separate "test" set.

Two major causes for poor generalization ability are overfitting and underfitting. Overfitting refers to that a model perform well on training data but poorly on test data, and underfitting refers to that a model neither performs well on training data nor on test data. Underfitting is easy to detect by using a good performance metric such as classification accuracy for classification problems. Underfitting occurs when (1) the features are not good for the learning task, or (2) the model is not complex enough to well fit the training data, e.g., fitting non-linear data with a linear model. To address overfitting in the second case, we usually choose a machine learning algorithm with more complex models. Overfitting occurs when the model is too complex and thus "memorize" the training data rather than learning patterns in the data for generalization. To address overfitting, we can either collect more training data or control the complexity of models.

Regularization is a kind of techniques that control model complexity through constraining or shrinking model parameters towards zero. Different regularization techniques have been studied, such as penalizing the L1 or L2 norm of the model parameters. For the multiclass logistic regression introduced in previous section, we can introduce L2 regularization into the loss function in Eq. (2.7) and get a regularized loss

$$L(f) = \frac{1}{|\mathcal{D}|} \sum_{(x,y)\in\mathcal{D}} l(y, f(x; W)) + \lambda ||W||^2, \tag{2.9}$$

where λ is a hyper-parameter determining how much attention we should pay to the regularization. Larger λ means we want to significantly control the complexity of the learnt model.

So far we only intuitively described the relation between model complexity and generalization ability. Formal analysis can be found in statistical learning theory [48, 49], which is a branch of machine learning.

2.4 Building a Machine Learning Model

In previous sections, we discussed different machine learning paradigms and the key components of a machine learning algorithm. In this section, we will discuss another important part of machine learning: how to build a machine learning algorithm to solve a practical problem.

Figure 2.1 shows a typical workflow for building a machine learning model in practice. We take spam email detection as an example to introduce the workflow.

2.4.1 Data Collection and Feature Engineering

To solve a real-world problem using machine learning, the first step we need to do is to collect data. For spam email detection, we need emails, including both normal emails and spam emails. To ensure the performance of final machine learning model, the emails we collected should have enough diversity and coverage, e.g., including emails in different languages, with different senders and receivers, different length, etc.

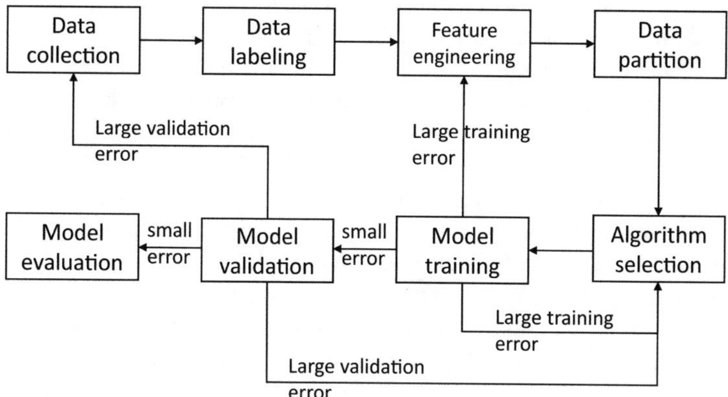

Fig. 2.1 Typical workflow for building a machine learning model

Then we need to label which emails are normal ones and which are spams. To ensure the quality of the labeling, usually an email is labeled by multiple human labelers and the final label is obtained by majority voting.

Next step is feature engineering, including feature extraction, optionally feature selection and dimension reduction, and feature normalization. Most machine learning algorithms such as logistic regression, support vector machines, and boosting trees cannot handle raw data like images, text, and audio. Thus, feature extraction is a crucial step to convert raw data into the format (e.g., integers or float numbers) that can be easily processed by machine learning algorithms.

For spam email detection, before feature extraction, we need to pre-process the text of emails, including removing punctuation and stop words, word stemming, and word lemmatization. Now we extract a set of features from cleaned text. Possible features include length of the email title/body, occurrences of certain words/phrases such as "advertise", "promotion", "discount", "free shipping", etc.

Some of the extracted features may be highly correlated and therefore redundant to a certain degree. Feature selection and dimension reduction techniques can help to compress the features into a lower dimensional space, which correspondingly saves storage space and speeds up the training.

Furthermore, for some machine learning algorithms, we need to normalize each feature into a fixed range such as $[0, 1]$ or $[-1, 1]$, so as to make the training stable and/or converge faster.

2.4.2 Algorithm Selection, Model Training, and Hyper-parameter Tuning

As aforementioned, the model learnt on training data will be applied to make predictions for unseen data. To ensure a model not only perform well on the training set but also generalize well to unseen data, a common practice in machine learning is to randomly partition the dataset into three disjoint subset: a training set, a validation set, and a test set. The training set is used to train and optimize machine learning models, the validation set is used to validate the performance of the trained models and choose the best hyper-parameters, and finally the chosen model is evaluated on the test set.

The model training and hyper-parameter tuning are usually conducted in an iterative manner. Suppose we choose the logistic regression algorithm for spam email detection. We first train a model $\mathcal{M}_1$ with hyper-parameter λ_1 and test this model on the validation set. Then we train another model $\mathcal{M}_2$ with a hyper-parameter $\lambda_2 > \lambda_1$ and test it on the validation set. If $\mathcal{M}_2$ is better than $\mathcal{M}_1$ on the validation set, then we will try a even larger hyper-parameter; if $\mathcal{M}_2$ is worse than $\mathcal{M}_1$, we will try a smaller $\lambda_3 < \lambda_1$. We continue this hyper-parameter tuning process until we are satisfied with the model performance on the validation set. Then we choose the model with the best accuracy on the validation set and test it on the test set.

Another possibility is that no matter how we tune the hyper-parameter, we do not get a model performing well on the validation set. For this situation, we need to diagnose the model performances on the training and validation sets:

- If the logistic regression algorithm performs poorly on both the datasets, this indicates that our models are not expressive enough to fit the training data, i.e., underfitting. We can have two solutions to address underfitting:
  - We can turn to algorithms with more complex models. For example, we can try deep neural networks or tree based learning algorithms for this classification problem, instead of the simple linear model in logistic regression. Since we change the algorithm, this is also called algorithm selection, by using the validation set.
  - We may go back to the feature engineering step and improve the expressiveness of the features. We can extract more features; we can also make existing features more powerful, e.g., increasing the granularity by using float values instead of integers.
- If the algorithm perform well on the training set but poorly on the validation set, this indicates that the model overfits the training data. To address overfitting,
  - We can go to the data collection step and collect more data, so that we have more data for training.
  - We can consider other regularization techniques so as to further limit model complexity.

References

1. Bahdanau, D., Cho, K., & Bengio, Y. (2015). Neural machine translation by jointly learning to align and translate. In *3rd International Conference on Learning Representations, ICLR 2015*.
2. Bishop, C. M. (2006). *Pattern recognition and machine learning*. springer.
3. Burges, C. J. C., et al. (2010). Dimension reduction: A guided tour. *Foundations and Trends® in Machine Learning, 2*(4), 275–365.
4. Cai, D., He, X., Li, Z., Ma, W.-Y., & Wen, J.-R. (2004). Hierarchical clustering of www image search results using visual, textual and link information. In *Proceedings of the 12th Annual ACM International Conference on Multimedia* (pp. 952–959).
5. Cao, Z., Qin, T., Liu, T.-Y., Tsai, M.-F., & Li, H. (2007). Learning to rank: from pairwise approach to listwise approach. In *Proceedings of the 24th international conference on Machine learning* (pp. 129–136).
6. Caruana, R. (1997). Multitask learning. *Machine Learning, 28*(1), 41–75.
7. Chen, T., & Guestrin, C. (2016). Xgboost: A scalable tree boosting system. In *Proceedings of the 22nd acm Sigkdd International Conference on Knowledge Discovery and Data Mining* (pp. 785–794).
8. Devlin, J., Chang, M.-W., Lee, K., & Toutanova, K. (2019). Bert: Pre-training of deep bidirectional transformers for language understanding. In *NAACL-HLT (1)*.

9. Duchi, J., Hazan, E., & Singer, Y. (2011). Adaptive subgradient methods for online learning and stochastic optimization. *Journal of Machine Learning Research, 12*(Jul), 2121–2159.
10. Fernando, B., Bilen, H., Gavves, E., & Gould, S. (2017). Self-supervised video representation learning with odd-one-out networks. In *Proceedings of the IEEE Conference on Computer Vision and Pattern Recognition* (pp. 3636–3645).
11. Fodor, I. K. (2002). A survey of dimension reduction techniques. Technical report, Lawrence Livermore National Lab., CA (US).
12. Freund, Y., Schapire, R., & Abe, N. (1999). A short introduction to boosting. *Journal-Japanese Society For Artificial Intelligence, 14*(771-780), 1612.
13. Friedman, J., Hastie, T., & Tibshirani, R. (2001). *The elements of statistical learning* (vol. 1). Springer series in statistics. New York: Springer.
14. Gao, B., Liu, T.-Y., Qin, T., Zheng, X., Cheng, Q.-S., & Ma, W.-Y. (2005). Web image clustering by consistent utilization of visual features and surrounding texts. In *Proceedings of the 13th Annual ACM International Conference on Multimedia* (pp. 112–121).
15. Gupta, V., & Lehal, G. S. (2010). A survey of text summarization extractive techniques. *Journal of Emerging Technologies in Web Intelligence, 2*(3), 258–268.
16. He, D., Xia, Y., Qin, T., Wang, L., Yu, N., Liu, T.-Y., et al. (2016). Dual learning for machine translation. In *Advances in Neural Information Processing Systems* (pp. 820–828).
17. Jain, A. K. (2010). Data clustering: 50 years beyond k-means. *Pattern Recognition Letters, 31*(8), 651–666.
18. Jain, A. K., Narasimha Murty, M., & Flynn, P. J. (1999). Data clustering: a review. *ACM Computing Surveys (CSUR), 31*(3), 264–323.
19. Karypis, M. S. G., Kumar, V., & Steinbach, M. (2000). A comparison of document clustering techniques. In *TextMining Workshop at KDD2000 (May 2000).*
20. Ke, G., Meng, Q., Finley, T., Wang, T., Chen, W., Ma, W., et al. (2017). Lightgbm: A highly efficient gradient boosting decision tree. In *Advances in Neural Information Processing Systems* (pp. 3146–3154).
21. Kingma, D. P., & Ba, J. (2014). Adam: A method for stochastic optimization. Preprint. arXiv:1412.6980.
22. Koehn, P. (2009). *Statistical machine translation*. Cambridge: Cambridge University Press.
23. Krizhevsky, A., Sutskever, I., & Hinton, G. E. (2012). Imagenet classification with deep convolutional neural networks. In *Advances in Neural Information Processing Systems* (pp. 1097–1105).
24. Lee, J. M., & Sonnhammer, E. L. L. (2003). Genomic gene clustering analysis of pathways in eukaryotes. *Genome Research, 13*(5), 875–882.
25. Li, J., Koyamada, S., Ye, Q., Liu, G., Wang, C., Yang, R., et al. (2020). Suphx: Mastering mahjong with deep reinforcement learning. Preprint. arXiv:2003.13590.
26. Liu, T.-Y. (2011). *Learning to rank for information retrieval*. Springer Science & Business Media.
27. Lowe, D. G. (1999). Object recognition from local scale-invariant features. In *Proceedings of the Seventh IEEE International Conference on Computer Vision* (vol. 2, pp. 1150–1157). IEEE.
28. Mitchell, T. M. (1997). *Machine learning*. McGraw Hill.
29. Murphy, K. P. (2012). *Machine learning: a probabilistic perspective*. MIT Press.
30. Nallapati, R., Zhou, B., dos Santos, C., Gulcehre, C., & Xiang, B. (2016). Abstractive text summarization using sequence-to-sequence rnns and beyond. In *Proceedings of The 20th SIGNLL Conference on Computational Natural Language Learning* (pp. 280–290).
31. Navigli, R., & Crisafulli, G. (2010). Inducing word senses to improve web search result clustering. In *Proceedings of the 2010 Conference on Empirical Methods in Natural Language Processing* (pp. 116–126). Association for Computational Linguistics.
32. Pan, S. J., & Yang, Q. (2009). A survey on transfer learning. *IEEE Transactions on Knowledge and Data Engineering, 22*(10), 1345–1359.
33. Peters, M. E., Neumann, M., Iyyer, M., Gardner, M., Clark, C., Lee, K., et al. (2018). Deep contextualized word representations. In *Proceedings of NAACL-HLT* (pp. 2227–2237).

34. Philbin, J., Chum, O., Isard, M., Sivic, J., & Zisserman, A. (2007). Object retrieval with large vocabularies and fast spatial matching. In *2007 IEEE Conference on Computer Vision and Pattern Recognition* (pp. 1–8). IEEE.
35. Platt, J. C. (1999). Fast training of support vector machines using sequential minimal optimization. In *Advances in Kernel Methods: Support Vector Learning* (pp. 185–208).
36. Radzikowski, K., Nowak, R., Wang, L., & Yoshie, O. (2019). Dual supervised learning for non-native speech recognition. *EURASIP Journal on Audio, Speech, and Music Processing, 2019*(1), 3.
37. Schölkopf, B., Smola, A. J., Bach, F., et al. (2002). *Learning with kernels: support vector machines, regularization, optimization, and beyond.*
38. Sebastiani, F. (2002). Machine learning in automated text categorization. *ACM Computing Surveys (CSUR), 34*(1), 1–47.
39. Seber, G. A. F., & Lee, A. J. (2012). *Linear regression analysis* (vol. 329). John Wiley & Sons.
40. Sermanet, P., Lynch, C., Chebotar, Y., Hsu, J., Jang, E., Schaal, S., et al. (2018). Time-contrastive networks: Self-supervised learning from video. In *2018 IEEE International Conference on Robotics and Automation (ICRA)* (pp. 1134–1141). IEEE.
41. Silver, D., Huang, A., Maddison, C. J., Guez, A., Sifre, L., Van Den Driessche, G., et al. (2016). Mastering the game of go with deep neural networks and tree search. *Nature, 529*(7587), 484.
42. Silver, D., Hubert, T., Schrittwieser, J., Antonoglou, I., Lai, M., Guez, A., et al. (2018). A general reinforcement learning algorithm that masters chess, shogi, and go through self-play. *Science, 362*(6419), 1140–1144.
43. Silver, D., Schrittwieser, J., Simonyan, K., Antonoglou, I., Huang, A., Guez, A., et al. (2017). Mastering the game of go without human knowledge. *Nature, 550*(7676), 354–359.
44. Steinbach, M., Tan, P.-N., Kumar, V., Klooster, S., & Potter, C. (2003). Discovery of climate indices using clustering. In *Proceedings of the Ninth ACM SIGKDD International Conference on Knowledge Discovery and Data Mining* (pp. 446–455).
45. Sutton, R. S., & Barto, A. G. (2018). *Reinforcement learning: An introduction.* MIT Press.
46. Thalamuthu, A., Mukhopadhyay, I., Zheng, X., & Tseng, G. C. (2006). Evaluation and comparison of gene clustering methods in microarray analysis. *Bioinformatics, 22*(19), 2405–2412.
47. Tung, H.-Y., Tung, H.-W., Yumer, E., & Fragkiadaki, K. (2017). Self-supervised learning of motion capture. In *Advances in Neural Information Processing Systems* (pp. 5236–5246).
48. Valiant, L. G. (1984). A theory of the learnable. *Communications of the ACM, 27*(11), 1134–1142.
49. Vapnik, V. (2013). *The nature of statistical learning theory.* Springer Science & Business Media.
50. Wang, Y., Xia, Y., Zhao, L., Bian, J., Qin, T., Liu, G., et al. (2018). Dual transfer learning for neural machine translation with marginal distribution regularization. In *Thirty-Second AAAI Conference on Artificial Intelligence.*
51. Weisberg, S. (2005). *Applied linear regression* (vol. 528). John Wiley & Sons.
52. Xing, Z., Pei, J., & Keogh, E. (2010). A brief survey on sequence classification. *ACM Sigkdd Explorations Newsletter, 12*(1), 40–48.
53. Zamir, O., & Etzioni, O. (1998). Web document clustering: A feasibility demonstration. In *Proceedings of the 21st Annual International ACM SIGIR Conference on Research and Development in Information Retrieval* (pp. 46–54).
54. Zeiler, M. D. (2012). Adadelta: an adaptive learning rate method. Preprint. arXiv:1212.5701.
55. Zhu, X., & Goldberg, A. B. (2009). Introduction to semi-supervised learning. *Synthesis Lectures on Artificial Intelligence and Machine Learning, 3*(1), 1–130.

Chapter 3
Deep Learning Basics

3.1 Neural Networks

Deep learning is not a new concept and has a long and rich history, under different names. It has become very popular and successful in last decade as large scale of training data, powerful computational hardware, and easy-to-use open-source software become available.

Deep learning is mainly based on artificial neural networks (ANNs). A neural network is a collection of nodes/neurons connected together. The earliest and simplest neural network is a single neuron, also called *perceptron*, which linearly combines multiple inputs, performs a nonlinear transformation, and outputs a scalar value, as shown in Fig. 3.1. Mathematically, a perceptron can be written as a function taking a vector x as input and outputting a scalar value y

$$y = g(\sum_{j=1}^{d} w_j x_j + b), \tag{3.1}$$

where x_j is the j-th dimension of x, b is the bias term, and $g()$ is an activation function, which enables that a perceptron can represent nonlinear functions. w_j and b are the parameters to be learnt from data.

There are different choices for the activation function. In original perceptron, the activation function g is a hard threshold function, i.e., $g(z) = 1, \forall z > 0$ and $g(z) = 0, \forall z \leq 0$. In sigmoid perceptron, g is a logistic function:

$$g(z) = \sigma(z) = \frac{1}{1 + e^{-z}}. \tag{3.2}$$

$\sigma(\sum_{j=1}^{d} w_j x_j + b)$ is usually interpreted as the probability of the sample x being in one class, and its probability of being in the other class is $1 - \sigma(\sum_{j=1}^{d} w_j x_j + b)$.

T. Qin, *Dual Learning*, https://doi.org/10.1007/978-981-15-8884-6_3

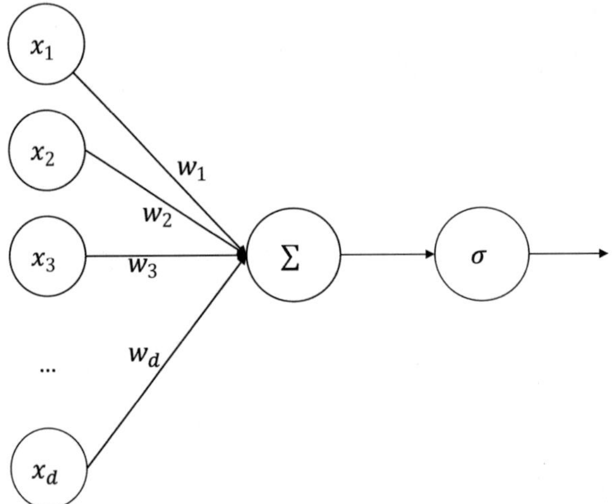

Fig. 3.1 The simplest neural network: perceptron, which was originally inspired by biological neural systems [48]

Multiple connected nodes/neurons[1] form a neural network. There two fundamentally distinct ways to connect neurons, resulting in two kinds of neural networks: *feedforward networks* and *recurrent networks*. Feedforward networks are the most basic neural networks. They are called feedforward because they only contain feedforward connections, and the information in a feedforward network flows in one direction: from the input x, through the hidden nodes, and finally to the output y. The nodes and connections in a feedforward network form a directed acyclic graph. In addition to feedforward connections, a recurrent network also contains feedback connections, through which the information from the output of the network can be fed back to its input or internal nodes. This section will concentrate on feedforward networks; recurrent networks will be introduced in Sect. 3.3.1.

Feedforward networks are usually organized in layers, and a feedforward network contains an input layer, an output layer, and (optionally) multiple middle layers between the input and output layers. Those middle layers are also called hidden layers. A perceptron is a feedforward network with only the input and output layers. In feedforward networks, a node in the hidden and output layers usually only receive input nodes in its immediately preceding layer.[2]

Figure 3.2 shows a 4-layer feedforward network including an input layer with 5 nodes (i.e., the input is a 5-dimensional vector), two hidden layers each with

[1]In this book, we will use nodes and neurons exchangeably when we are talking about neural networks.

[2]In more recent neural networks, a node may receive input from multiple layers. We will discuss those networks in Sects. 3.2 and 3.3.3.

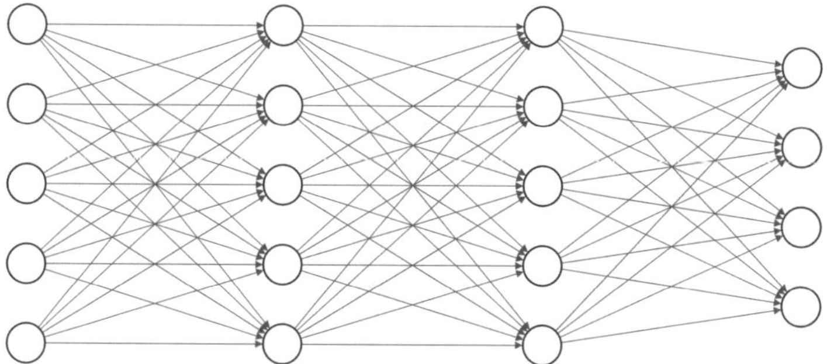

Fig. 3.2 A feedforward neural network with 4 layers: an input layer, two hidden layers, and an output layer

5 nodes, and an output layer with 4 nodes (i.e., this network can be used for 4-class classification). In this network, two adjacent layers are fully connected. This kind of fully connected multilayer feedforward networks is also called multilayer perceptrons.

As can be seen from the figure, each layer of a feedforward neural network consists of a set of neurons. We usually use a vector to represent the values of the neurons in a layer. Each neuron (except those input neurons) together with its input neurons forms a perceptron. Now we consider an l-layer feedforward network with i-th layer containing d_i nodes. Suppose the first layer is the input layer and the l-th layer is the output layer. Let $x \in R^{d_1}$ denote the input vector of a sample, $y \in R^{d_l}$ the output of a feedforward network with input x, $h^{(i)} \in R^{d_i}$ the values of the i-th layer for $i \in \{2, \cdots, l-1\}$, $b^{(i)}$ the bias vector associated with the i-th layer for $i \in \{2, \cdots, l\}$ and $W^{(i)}$ denote the weight/parameter matrix between the $(i-1)$-th and i-th layer for $i \in \{2, \cdots, l\}$, in which $W^{(i)}_{j,k}$ is the weight/parameter of the edge connecting the j-th node in the $(i-1)$-th layer to the k-th node in the i-th layer. Mathematically, we have

$$h^{(i)} = g^{(i)}(W^{(i)}h^{(i-1)} + b^{(i)}), \forall i \in \{2, \cdots, l-1\},$$

and

$$y = W^{(l)}h^{(l-1)} + b^{(i)}),$$

where $g^{(i)}$ is element-wise activation function for layer $i \in \{2, \cdots, l-1\}$.

Sigmoid function (Eq. (3.3)) and TanH function (Eq. (3.4)) have been used as activation functions in neural networks in early days.

$$g(z) = \sigma(z) = \frac{1}{1 + e^{-z}} \tag{3.3}$$

$$g(z) = \tanh(z) = \frac{e^z - e^{-z}}{e^z + e^{-z}} \tag{3.4}$$

Rectified linear unit (ReLU) (Eq. (3.5)) [38] and its variants such as leaky ReLU (Eq. (3.6)) [36] and parametric leaky ReLU (Eq. (3.7)) [16] are more popular activation functions in deep neural networks.

$$g(z) = \begin{cases} 0 & \text{if } z \leq 0 \\ z & \text{if } z > 0 \end{cases} \tag{3.5}$$

$$g(z) = \begin{cases} 0.01z & \text{if } z \leq 0 \\ z & \text{if } z > 0 \end{cases} \tag{3.6}$$

$$g(z) = \begin{cases} \alpha z & \text{if } z \leq 0 \\ z & \text{if } z > 0, \end{cases} \tag{3.7}$$

where α is a parameter to be learnt during the training of neural networks.

3.2 Convolutional Neural Networks

Multilayer perceptrons are fully connected feedforward networks, i.e., each neuron in one layer is connected to all the neurons in the next layer. The full connection results in a large number of parameters, and makes the networks computationally costly and prone to overfitting training data. Suppose we need to classify images with size 1000×1000. If we use a multilayer perceptron with a hidden layer containing 10000 neurons, there will be 10 billion parameters ($1000 \times 1000 \times 10000$) connecting the input layer and the hidden layer! Clearly multilayer perceptrons with a huge number of parameters are computationally unaffordable and easy to overfit. They are not a good choice to handle high dimensional 2D data like images.

Convolutional neural networks (CNNs) are a specific kind of feedforward neural networks, which are well known for and named after the convolution operation. Thanks for the convolution operation, CNNs can efficiently process 2D grid-like data such as images and videos in computer vision tasks.[3]

Inspired by biological processes [24], convolution leverages two important ideas that can significantly improve computational efficiency of a network: sparse connection and parameter sharing. A neuron in a convolutional layer only connects to a few neurons (also called receptive field) in its previous layer, and the weights of those connections are shared across all the neurons in the convolutional layer.

[3]CNNs have been extended to process sequence data such as time series in financial applications and text sequences in natural language tasks [13].

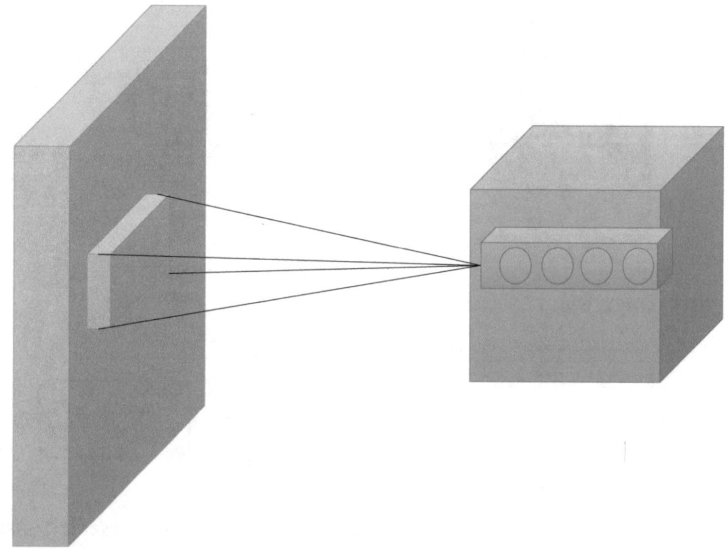

Fig. 3.3 A convolutional layer with 3D convolutional kernels

Those shared weights compose a convolution kernel or filter. Let us consider a filter that has a 10×10 receptive field, i.e., a neuron in the convolutional layer connects to 100 neurons in the previous layer. When this filter convolves an image with size 1000×1000 and suppose there is no overlap among the convolutions, we get 10000 neurons in the convolutional layer. That is, the convolution operation only needs 101 parameters[4] to map the input layer with 1000×1000 neurons to a layer with 10000 neurons, which is significantly more efficient than multilayer perceptrons.

A color image is a 3D tensor with width, height and depth. We can use 3D convolutional filters to process color images. If a convolutional layer contains multiple convolutional filters, the output of this convolutional layer is also a 3D tensor, as shown in Fig. 3.3.

Pooling, a kind of nonlinear down-sampling, is another important operation in CNNs. As shown in Fig. 3.4, pooling partitions an input image into a set of non-overlapping rectangles, and for each such sub-region outputs the maximum (known as max pooling).[5] Through down-sampling, pooling can reduce the spatial size of the representation, the number of parameters (and hence also reduce overfitting), memory footprint, and computational cost.

CNNs are composed by convolutional layers, pooling layers, and fully connected layers, as illustrated in Fig. 3.5. As can be seen, both convolutional layers and fully connected layers contain the ReLU activation, and M/N/K indicate the repeat of the

[4] 100 weights for the connections plus a bias term.

[5] There are other kinds of pooling operations, such as average pooling or L2 pooling.

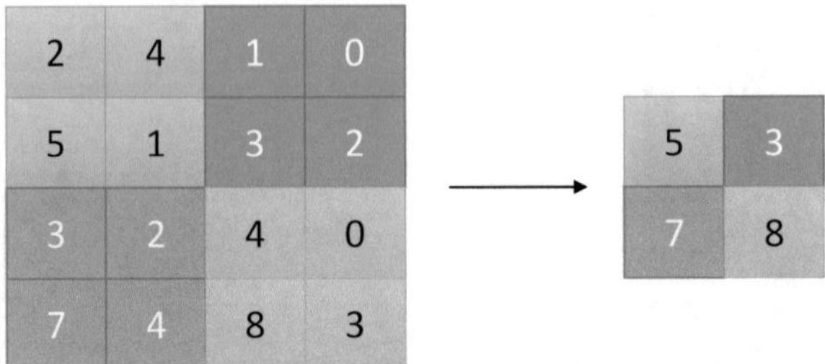

Fig. 3.4 Max pooling with a 2×2 filter

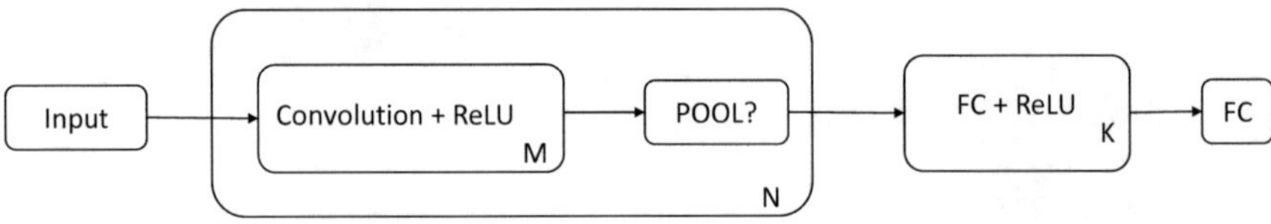

Fig. 3.5 Abstraction of CNN architectures. "POOL?" means that a pooling layer is optional after every M convolutional layers

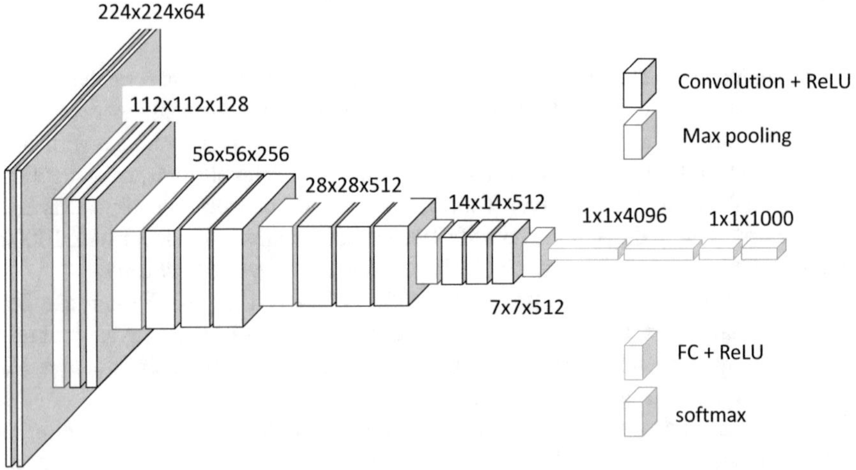

Fig. 3.6 The 16-layer VGG. Here we do not count layers without learnable parameters, such as the input layer and pooling layers

corresponding block for M/N/K times. Figure 3.6 shows a real example of CNNs, the 16-layer VGG network [51].

In addition to the basic convolution and pooling operations, new operations and modules have been introduced to further improve CNNs for computer visions tasks. For example, the inception module is introduced in Inception networks [55] to improve computational efficiency of CNNs, residual connections are introduced in ResNet [17] to train very deep CNNs (up to 1000 layers), and densely connected layers are introduced in DenseNet [22] to encourage feature reuse and reduce the number of parameters.

3.3 Sequence Modeling

Sequential data is very common in many applications such as machine translation, text classification, speech recognition, time series predictions, etc. Different neural networks have been designed to process sequential data, including recurrent neural networks (RNNs) and latest Transformer networks.

3.3.1 Recurrent Neural Network and Its Variants

The main challenge to handle sequential data is that the length of a sequence can vary a lot. It is difficult for feedforward neural networks to process sequential data. To handle sequences with variable lengths, RNNs take advantage of *parameter sharing* and process the inputs at different positions of a sequence with the same set of parameters. Let's take sentence classification as an example and consider a sentences with n words $(x^{(1)}, x^{(2)}, \cdots, x^{(n)})$, where $x^{(t)}$ is the word at the t-th position of the sentence. An RNN computes a hidden state $h^{(t)}$ at each position t through a recurrent function based on current word $x^{(t)}$ and previous hidden state $h^{(t-1)}$

$$h^{(t)} = f(h^{(t-1)}, x^{(t)}; \theta), \tag{3.8}$$

where θ is the parameters of the RNN. As can be seen, θ is independent to position t and shared across all the positions in a sentence. Therefore, no matter how long a sentence is, it can be handled by an RNN with a fixed set of parameters. Unfolding this equation, we get

$$h^{(t)} = f(f(f(h^{(t-3)}, x^{(t-2)}; \theta), x^{(t-1)}; \theta), x^{(t)}; \theta)$$

$$= f(f(f(\cdots f(f(f(h^{(0)}, x^{(1)}; \theta), x^{(2)}; \theta), x^{(3)}; \theta), \cdots, x^{(t-1)}; \theta), x^{(t)}; \theta).$$

We can see that $h^{(t)}$ encodes the information of the words up to position t, and similarly $h^{(n)}$ encodes the information of the whole sentence.

A simple implementation of the above recurrent function is

$$h^{(t)} = f(h^{(t-1)}, x^{(t)}; \theta) = \sigma(W_{hh}h^{(t-1)} + W_{xh}x^{(t)} + b),$$

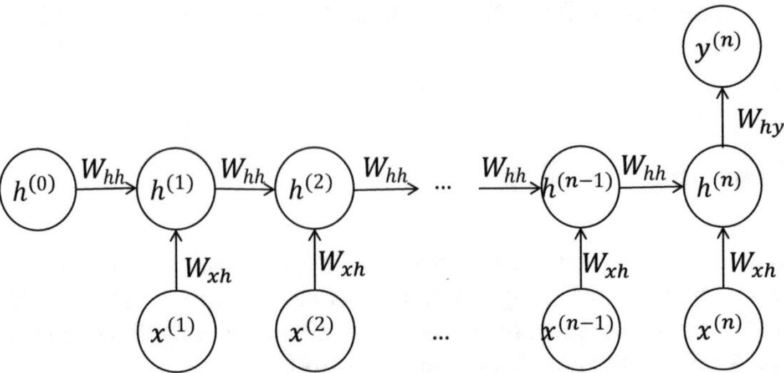

Fig. 3.7 A recurrent neural network for sentence/sequence classification. All the nodes in the figure are vectors. We can simply set $h^{(0)}$ to a zero vector. Note that we omit the bias vectors b and c from the figure for simplicity. For sentence classification in natural language processing, $x^{(i)}$ is the embedding vector of the i-th word in the sentence

where $\theta = (W_{hh}, W_{xh}, b)$, b is a bias vector, $\sigma()$ is the activation function, W_{xh} is the weight matrix associated with the input to hidden connections, and W_{hh} is the weight matrix associated with hidden-to-hidden recurrent connections.

For sentence classification tasks, as shown in Fig. 3.7, we use the last hidden state $h^{(n)}$ as the representation of the whole sentence and connect it to an output node $y^{(n)}$ with another weight matrix W_{hy}:

$$y^{(n)} = W_{hy} y^{(n)} + c, \tag{3.9}$$

where c is a bias vector. One can use the $softmax()$ function to convert the output $y^{(n)}$ to a probability distribution.

For sentence classification tasks, we only need to predict a category label for a sentence. For some sentence modeling tasks such as language modeling and part-of-speech tagging, as shown in Fig. 3.8, we need to predict a label for each word/token in a sentence, which is implemented by connecting each hidden state to an output node:

$$y^{(t)} = W_{hy} y^{(t)} + c.$$

Figures 3.7 and 3.8 illustrate RNNs with only hidden layer for different tasks. We can get deep RNNs by adding more hidden layers. Figure 3.9 shows an RNN with two hidden layers.

Training an RNN is difficult, especially for long sentences/sequences, because the gradients propagated over many steps tend to either vanish (when the maximal absolute eigenvalue of W_{hh} is smaller than and not close to 1) or explode (when the maximal absolute eigenvalue of W_{hh} is larger than and not close to 1). This difficulty is called the challenge of long-term dependencies. To address this challenge,

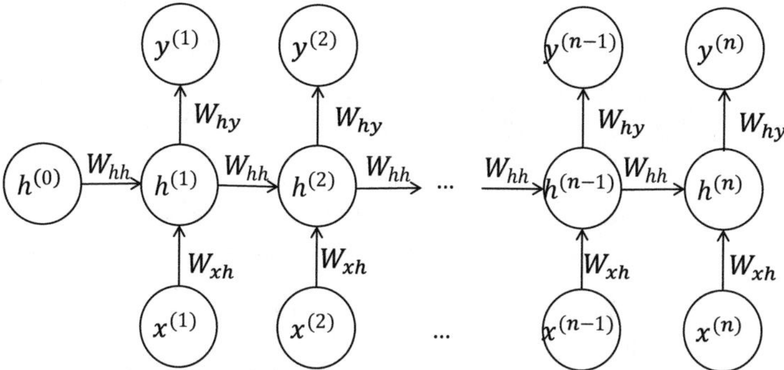

Fig. 3.8 A recurrent neural network that predicts a label for each word/token of a sequence. All the nodes in the figure are vectors. We can simply set $h^{(0)}$ to a zero vector. Note that we omit the bias vectors b and c from the figure for simplicity

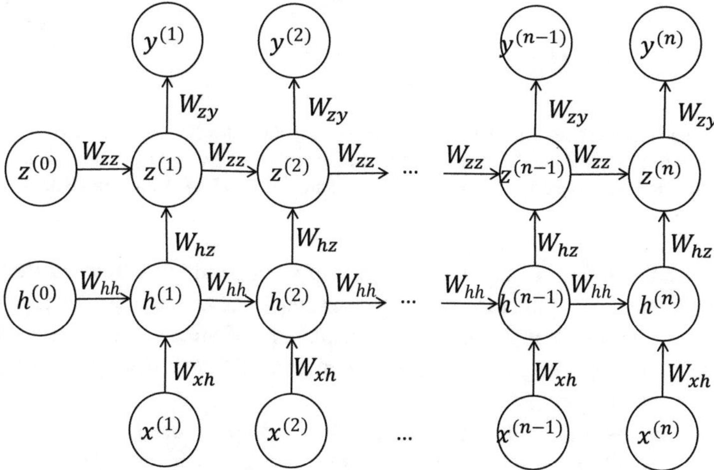

Fig. 3.9 A recurrent neural network with two hidden layers

multiple variants of RNNs have been introduced, and Long Short Term Memory (LSTM) [19] and Gated Recurrent Unit [3] are most popular among those variants.

3.3.2 *Encoder-Decoder Architecture*

Sequence to sequence (seq2seq) learning is a specific kind task of sequence modeling, which takes a sequence as input and output another sequence. It covers many real world problems, including machine translation, text summarization, question answering, etc. A key difference between those problems and previous

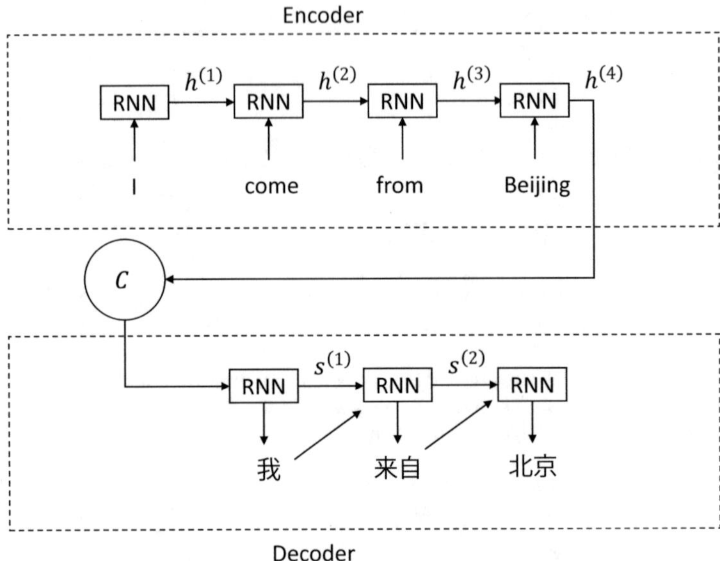

Fig. 3.10 The basic encoder-decoder architecture for machine translation. Here we use $h^{()}$ and $s^{()}$ to denote the hidden states in the encoder and decoder respectively. The RNN units in the encoder and decoder can be replaced by any other RNN variants, such as LSTM units or GRU units

part-of-speech tagging is that in seq2seq the input and output sequences are usually of different lengths and the tokens in the input sequence and output sequence are not well aligned, while the input and output sequences of are the same length and the i-th token in the input sequence is aligned with the i-th token in the output sequence in part-of-speech tagging.

The encoder-decoder architecture [3, 54] is widely used and the dominant approach for seq2seq learning. Figure 3.10 explains the basic idea of the architecture taking machine translation as an example. As can be seen, the encoder-decoder architecture has two major components:

- An encoder RNN[6] encodes the input sequence into a fix-sized vector ($h^{(4)}$ in the figure), which is then processed and passed to the decoder as context (C in the figure).
- An decoder RNN takes the context as input and generates a sequence of words one by one, from left to right.

Mathematically, we have

$$h^{(t)} = RNN(h^{(t-1)}, x^{(t)}; \theta_{en})$$

[6]A RNN contains many RNN units and all those units share parameters.

for the encoder and

$$s^{(t)} = RNN(s^{(t-1)}, y^{(t)}; \theta_{de})$$

for the decoder, where θ_{en} and θ_{de} are the parameters of the encoder RNN and decoder RNN respectively. Note that $s^{(0)} = C$ and usually we set $h^{(0)}$ to a zero vector.

An obvious limitation of the basic encoder-decoder architecture is that any sentence is encoded into a fixed-size representation and the decoder only takes this representation as input. Unfortunately, using a fixed-size representation to capture all the semantics of a very long or complex sentence is difficult, as demonstrated in [3, 54]. To address this limitation, the attention mechanism is introduced into the encoder-decoder architecture [1].

As shown in Fig. 3.11, the generation of the hidden state $s^{(i)}$ at position i of the decoder conditions on three pieces of information:

$$s^{(i)} = RNN(s^{(i-1)}, y^{(i-1)}, C^{(i)}),$$

where $C^{(i)}$ is a dedicated context vector for position i and generated by an attention module through a linear combination of the hidden representations from the encoder:

$$C^{(i)} = \sum_j \alpha_{i,j} h^{(j)}.$$

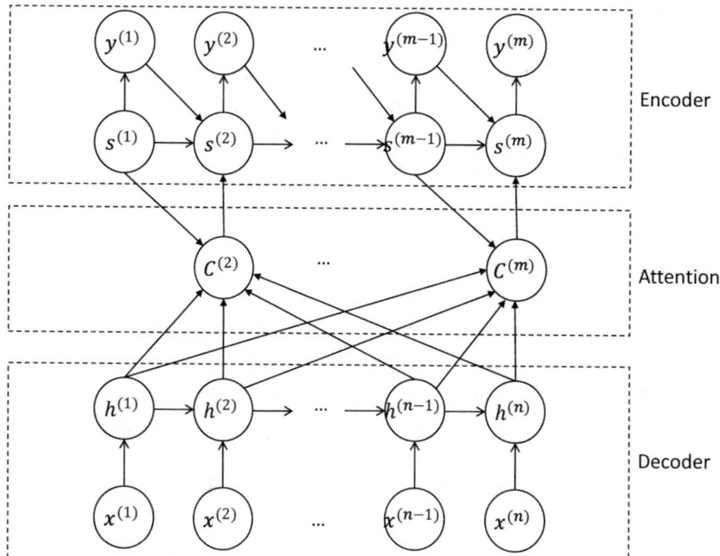

Fig. 3.11 The encoder-decoder-attention architecture. Here $C^{(i)}$ denote the context vector dedicated to position i in the decoder

The $\alpha_{i,j}$ is the attention weight denoting how much attention the decoder hidden state $s^{(i)}$ should put on the encoder hidden step $h^{(j)}$. There are multiple ways to compute the attention weights. A popular choice is

$$\alpha_{i,j} = softmax_j(\beta_{i,j}) = \frac{\exp(\beta_{i,j})}{\sum_j \exp(\beta_{i,j})},$$

where

$$\beta_{i,j} = \left(s^{(i-1)}\right)^T h^{(j)}.$$

Considering two decoding positions i and k, since $s^{(i-1)}$ is usually different from $s^{(k-1)}$, $\beta_{i,j}$ and $\alpha_{i,j}$ will also be different from $\beta_{k,j}$ and $\alpha_{k,j}$. Therefore, while generating hidden states $s^{(i)}$ (and so $y^{(i)}$) at different positions, the decoder can focus on and attend to different parts and words (represented by $h^{(j)}$) of the input sentence. That is, the attention module generates a different context for each decoder step and thus helps to pass richer information from the encoder to the decoder.

3.3.3 Transformer Networks

Transformer [58], the state-of-the-art encoder-decoder architecture, was first proposed for neural machine translation and other language generation tasks, and soon extended to many other domains, including pre-training [7, 45, 52], computer vision [14, 43], speech processing [8, 47], and music creation [21], etc.

Transformer gives up the recurrent units in RNNs and fully embraces the attention mechanism. Figure 3.12 shows the model architecture of Transformer.

The encoder is composed of a stack of N identical layers (but with different parameters), and each layer is composed of two sub-layers: a multi-head self-attention sub-layer, and a simple, position-wise fully connected feed-forward network. It employs a residual connection around each of the two sub-layers, followed by layer normalization.

The overall structure of the decoder is very similar to the encoder: a stack of N identical layers and with two sub-layers in each layer. There are two differences between the encoder and decoder. First, the decoder inserts a third sub-layer, which performs multi-head attention over the final output of the encoder. Second, the decoder adds a mask into the self-attention sub-layer so that the generation of hidden states and the final output at position i only depends on its preceding positions but not future positions.

The key innovation of Transformer is the multi-head self-attention mechanism. We first introduce self-attention and then multi-head self-attention.

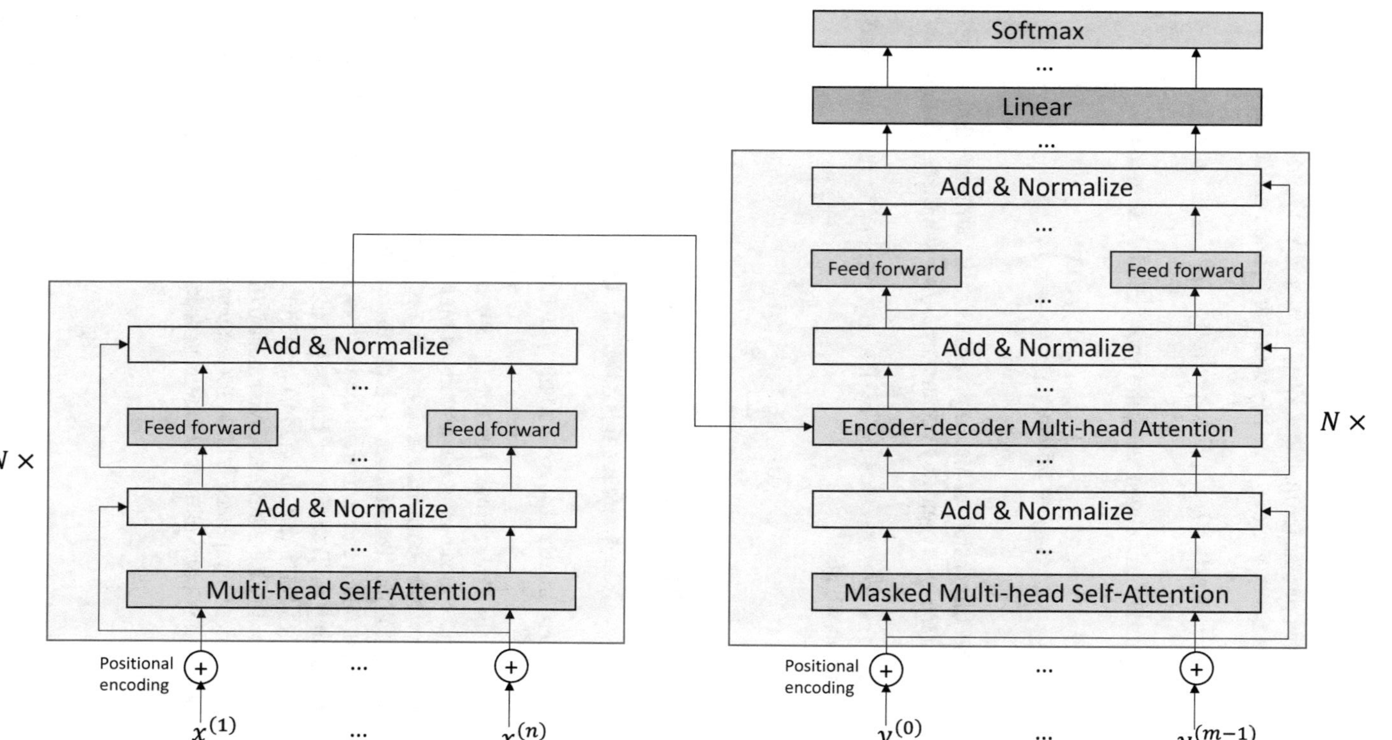

Fig. 3.12 Model architecture of Transformer

The self-attention function takes a sequence of vectors $a^{(1)}, a^{(2)}, \cdots, a^{(n)}$ as input and output another sequence of vectors $b^{(1)}, b^{(2)}, \cdots, b^{(n)}$.

1. It introduces three parameter matrices W^Q, W^K, and W^V and maps each input vector $a^{(i)}$ into a query vector $q^{(i)}$, a key vector $k^{(i)}$, and a value vector $v^{(i)}$:

$$q^{(i)} = W^Q a^{(i)}, \quad k^{(i)} = W^K a^{(i)}, \quad v^{(i)} = W^V a^{(i)}.$$

2. Attention weights are calculated using the query and key vectors:

$$\alpha_{i,j} = softmax_j \left(\frac{q^{(i)} \cdot k^{(j)}}{\sqrt{d_K}} \right),$$

   where d_K is the dimension of the key vectors. The normalization of the dot product between the query and key vectors over $\sqrt{d_K}$ stabilizes gradients during training.
3. The output vectors $b^{(i)}$ are obtained by linearly combining the values vectors with the attention weights:

$$b^{(i)} = \sum_j \alpha_{i,j} v^{(j)}.$$

For simplicity, we denote the self-attention function as

$$B = Attention(W^Q A, W^K A, W^V A),$$

where the i-column of matrix A is $a^{(i)}$ and the j-th column of matrix B is $b^{(j)}$.

One set of (W^Q, W^K, W^V) matrices is called an attention head. Each self-attention sub-layer in Transformer has multiple heads with different parameters (W_h^Q, W_h^K, W_h^V) and concatenates the outputs of multiple heads.

Transformer networks outperform RNNs in multiple aspects. First, without the recurrent operation, Transformer avoids the problem of gradient vanishing/explosion and is much easier to train than RNNs. Second, without the recurrent operation, the training of Transformer can be parallelized and its training is much more efficient than RNNs. Third, for models with similar numbers of parameters, the accuracy of Transformer is better than RNNs. Due to those advantages, Transformer networks have gradually replaced RNNs for sequence modeling and become the dominant model architecture in natural language tasks.

3.4 Training Deep Models

Similar to general machine learning algorithms, the training of deep neural networks is to find a set of network parameters θ through minimizing a loss function (or maximizing a reward function) $J()$ defined on a set D of training samples:

$$J(\theta) = \frac{1}{|D|} \sum_{(x,y)\in D} l(f(x;\theta), y), \tag{3.10}$$

where $f(;\theta)$ is the network specified by the parameters θ and $l(f(x;\theta), y)$ is the loss defined on a data pair with input x and desired output y.

We are facing several challenges while minimizing the above $J(\theta)$.

- First, the training data is usually of large scale, e.g., millions or even billions of sample as shown in Table 1.1. Therefore, gradient descent methods are inefficient and unaffordable.
- Second, the loss $J(\theta)$ is highly non-convex since deep neural networks are highly nonlinear. Consequently it is difficult (if not impossible) to find a global minimum.
- Third, since deep networks can have up to millions or even billions of parameters which are usually many more than the number of training samples, it is easy to overfit the training data, i.e., the learnt model works well on the training data but poorly on new samples in testing.

3.4.1 *Stochastic Gradient Descent*

To address the first challenge, in deep learning, stochastic gradient descent (SGD) and its variants are widely adopted. Rather than following the gradient of the entire training set in gradient descent, in SGD, we follow the gradient of a randomly selected minibatch of data for each iteration. The details of SGD is shown in Algorithm 1.

Algorithm 1 The stochastic gradient descent algorithm

Require: Learning rates $\gamma_1, \gamma_2, \cdots$
Require: Initial model parameters θ.
$t = 1$
repeat
  Sample a minibatch B_t of m training samples
  Compute the gradient over the minibatch $g_t = \frac{1}{m}\nabla_\theta \sum_{(x,y)\in B_t} l(f(x;\theta), y)$
  Update the model $\theta \leftarrow \theta - \gamma_t g_t$
  $t = t + 1$
until convergence

In general, to ensure the convergence (to a local minimum) of the SGD algorithm, we should gradually decrease the learning rate γ_t over time. A sufficient condition to guarantee the convergence of SGD is

$$\sum_{t=1}^{\infty} \gamma_t = \infty,$$

and

$$\sum_{t=1}^{\infty} \gamma_t^2 < \infty.$$

In practice, there are different ways to schedule the learning rates:

- One can decrease the learning rates for every n iterations, e.g., multiplying the learning rate by 0.5 for severy 100 minibatches.
- One can decrease the learning rates if the validation loss does not decrease for a certain number of minibatches, e.g., multiplying the learning rate by 0.1 if the validation loss does not decrease for 10 minibatches.
- One can linearly decrease the learning rates: $\gamma_t = (1-\frac{t}{\tau})\gamma_0+\frac{t}{\tau}\gamma_\tau$. After iteration τ, the learning rate is fixed to γ_τ.
- One can exponentially decrease the learning rates $\gamma_t = \gamma_0 \exp^{-kt}$, where k is a constant.

Many variants of SGD have been proposed to improve SGD, including momentum SGD [44], Nesterov accelerated SGD [39], AdaGrad [10], AdaDelta [63], Adam [29], etc. Which optimizer to choose depends on both the tasks and the architectures of neural networks.

3.4.2 Regularization

To address the challenge of overfitting, many regularization strategies have been designed and used in deep learning, including early stopping, data augmentation, dropout, limiting the model capacity through penalizing the norm of model parameters, etc.

When overfitting happens, we observe that the training loss decreases steadily over time, but the validation loss begins to increase at a certain time step. Obviously, we can obtain a better model with smaller validation loss at this time step. That is, even if the training loss is still decreasing, we stop the training process at this step according to the validation loss; such a strategy is known as *early stopping*.

The most natural way to avoid overfitting and make a learnt model generalize better is to train it with more data. Since collecting labeled data is often costly and time consuming, the training data we can have is always limited. Data augmentation is an effective approach to increase training data by creating fake data from the original training data or additional unlabeled data. Data augmentation methods can be very different depending on specific machine learning tasks.

- In computer vision tasks, the training data are often augmented with fake images generated from original training images through random rotation, resizing, vertical or horizontal flipping, cropping, color shifting/whitening [5, 31, 32, 35], and convex combinations of pairs of training images [18, 25, 64], while the labels of created fake images come from the original images.
- Data noising, which injects noise into training data by replacing words [30] or embeddings [12], masking words, flipping words or partial sentences [9], is a general data augmentation technique in natural language processing tasks [61].

Dropout [53] is a simple, computationally efficient yet effective regularization method for training deep neural networks. For each training sample or minibatch, it randomly selects neurons (hidden and visible) in a neural network, the contribution of those neurons to downstream neurons is temporally removed from the forward pass, and any weight updates are not applied to those dropped neurons in the backward pass. When neurons are randomly dropped out from a network during training, other neurons will have to step in and replace the role of the dropped neurons to make correct predictions. This reduces the mutual dependence among neurons. Dropout can be thought of as the ensembles of many neural networks: during training, dropout samples from an exponential number (2^n for a network with n neurons) of different "thinned" networks; at test time, it approximates the effect of averaging the predictions of all these thinned networks by simply using a single unthinned network that has smaller weights. This ensemble of a huge scale of thinned neural networks significantly reduces overfitting. DropConnect [59] extends Dropout from randomly dropping neurons to dropping weights in training.

Penalizing parameter norm is a widely used strategy in traditional machine learning. To penalize parameter norm in deep learning, we add a new term $\Omega(\theta)$ of the norm of model parameters into the loss defined in Eq. (7.2) and get a regularized loss $\hat{J}(\theta)$ to minimize:

$$\hat{J}(\theta) = J(\theta) + \alpha\Omega(\theta) = \frac{1}{|D|} \sum_{(x,y)\in D} l(f(x;\theta), y) + \alpha\Omega(\theta), \tag{3.11}$$

where α is a hyper-parameter to trade off the two loss terms. Different choices of the norm Ω can be used and will result in different solutions.

L2 norm, which is also know as weight decay, is one of the simplest and most widely used parameter norm penalty:

$$\Omega(\theta) = \frac{1}{2}\theta^T\theta = \frac{1}{2}\sum_i \theta_i^2,$$

where we assume the model parameters θ are in a vector form and θ_i is one dimension of the vector. To minimize the L2 norm regularized loss, we also need to minimize the norm of model parameters. That is, through L2 norm regularization, we prefer models with smaller norm and thus smaller model capacity, which is a straightforward approach to handle overfitting.

L1 norm, which is the sum of absolute values of the individual parameters, is also widely used for norm penalty:

$$\Omega(\theta) = ||\theta||_1 = \sum_i |\theta_i|.$$

L1 norm regularization [57] means that we prefer model sparsity and want to learn a model with as few non-zero weights as possible.

3.5 Why Deep Networks?

While artificial neural networks can be dated back to the study on perceptron more than 60 years ago [48], they gain renewed attention and become well recognized in this decade due to their great empirical/practical success in different areas including computer vision [17, 31], speech processing [47, 62], natural language processing [7, 15], and game playing [33, 50]. The recent success of neural networks should be largely accredited to their increased layers, i.e., deep neural networks. Figure 3.13 shows the error rate of neural networks on the ImageNet dataset (the ILSVRC challenge) and their number of layers. As can be seen, the accuracy of image classification on this large scale dataset is strongly and positively correlated

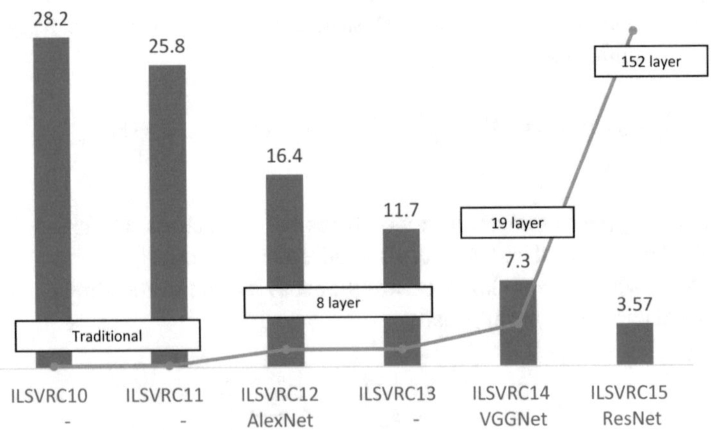

Fig. 3.13 Error rate on ImageNet versus network layers. Reproduced from https://sqlml.azurewebsites.net/2017/09/12/convolutional-neural-network/

with the depth of neural networks: the introduction of the first deep neural network, 8-layer AlexNet [31], in 2012 significantly improved the accuracy, reducing the error rate of traditional shallow models from 25.8% to 16.4%, and the 152-layer ResNet [17] introduced in 2015 further reduced the error rate to 3.57%, surpassing human error rate of 5.1%.

On one hand, deep neural networks have achieved great empirical success; on the other hand, researchers have also tried to theoretically understand their success from several aspects.

A line of works studies the expressiveness of deep neural networks and shows that they are more expressive and representationally powerful than shallow ones. Early studies show the universal approximation property of neural networks in approximations of different function classes [2, 20, 23, 42, 60] and the universal approximation theorem states that a feed-forward network with a single hidden layer (plus an input layer and an output layer) containing a finite number of neurons can well approximate real-valued continuous functions on compact subsets of R^n, under mild assumptions on the activation function. While those works show that arbitrarily small approximation error can be achieved if the network size is sufficiently large, they cannot explain the success of deep neural networks, i.e., why deep networks are better than shallow ones. Recent studies find that shallow networks need an exponential number of neurons to approximate certain functions, and their deep counterparts are more expressive and need substantially fewer neurons (e.g., a polynomial number of neurons) to approximate [6, 11, 34, 37, 46, 56].

It is well known that deep neural networks are highly non-convex and high dimensional, and finding a global minimum of a general non-convex function is NP-hard. Thus, while increasing the depth of a neural network make it more expressive, it also makes the optimization more difficult. Another line of works try to theoretically understand the advantages of deep neural networks from the perspective of optimization and shows that well-designed deep neural networks with specific architectures have better optimization properties than shallow ones. Several studies prove the existence of desirable loss landscape structures of deep neural networks with respect to a global minimum (e.g., all local minima are close to being globally optimal) under the strong assumptions of the model simplifications [4, 26] or of significant over-parameterization, e.g., the number of neurons of one hidden layer of a network is larger than the number of training samples [40, 41]. Several studies further show that under practical conditions that a popular and widely used deep neural network, deep ResNet, has no local minimum with a value higher than the global minimum value of corresponding scalar-valued [49] or vector-valued [27] basis-function models. Without any simplification assumption (e.g., model simplifications or over-parameterization), for deep nonlinear neural networks with the squared loss, Kawaguchi et al. [28] theoretically show that the quality of local minima tends to improve toward the global minimum value as depth and width increase.

References

1. Bahdanau, D., Cho, K., & Bengio, Y. (2015). Neural machine translation by jointly learning to align and translate. In *3rd International Conference on Learning Representations, ICLR 2015.*
2. Barron, A. R. (1993). Universal approximation bounds for superpositions of a sigmoidal function. *IEEE Transactions on Information Theory, 39*(3), 930–945.
3. Cho, K., van Merrienboer, B., Bahdanau, D., & Bengio, Y. (2014). On the properties of neural machine translation: Encoder-decoder approaches. In *Eighth Workshop on Syntax, Semantics and Structure in Statistical Translation (SSST-8), 2014.*
4. Choromanska, A., Henaff, M., Mathieu, M., Arous, G. B., & LeCun, Y. (2015). The loss surfaces of multilayer networks. In *Artificial Intelligence and Statistics* (pp. 192–204).
5. Cubuk, E. D., Zoph, B., Mane, D., Vasudevan, V., & Le, Q. V. (2019). Autoaugment: Learning augmentation strategies from data. In *Proceedings of the IEEE Conference on Computer Vision and Pattern Recognition* (pp. 113–123).
6. Delalleau, O., & Bengio, Y. (2011). Shallow vs. deep sum-product networks. In *Advances in Neural Information Processing Systems* (pp. 666–674).
7. Devlin, J., Chang, M.-W., Lee, K., & Toutanova, K. (2019). Bert: Pre-training of deep bidirectional transformers for language understanding. In *NAACL-HLT (1).*
8. Dong, L., Xu, S., & Xu, B. (2018). Speech-transformer: a no-recurrence sequence-to-sequence model for speech recognition. In *2018 IEEE International Conference on Acoustics, Speech and Signal Processing (ICASSP)* (pp. 5884–5888). IEEE.
9. Du, W., & Black, A. W. (2018). Data augmentation for neural online chats response selection. In *Proceedings of the 2018 EMNLP Workshop SCAI: The 2nd International Workshop on Search-Oriented Conversational AI* (pp. 52–58).
10. Duchi, J., Hazan, E., & Singer, Y. (2011). Adaptive subgradient methods for online learning and stochastic optimization. *Journal of Machine Learning Research, 12*(Jul), 2121–2159.
11. Eldan, R., & Shamir, O. (2016). The power of depth for feedforward neural networks. In *Conference on Learning Theory* (pp. 907–940).
12. Gao, F., Zhu, J., Wu, L., Xia, Y., Qin, T., Cheng, X., et al. (2019). Soft contextual data augmentation for neural machine translation. In *Proceedings of the 57th Annual Meeting of the Association for Computational Linguistics* (pp. 5539–5544).
13. Gehring, J., Auli, M., Grangier, D., Yarats, D., & Dauphin, Y. N. (2017). Convolutional sequence to sequence learning. In *Proceedings of the 34th International Conference on Machine Learning-Volume 70* (pp. 1243–1252). JMLR.org.
14. Girdhar, R., Carreira, J., Doersch, C., & Zisserman, A. (2019). Video action transformer network. In *Proceedings of the IEEE Conference on Computer Vision and Pattern Recognition* (pp. 244–253).
15. Hassan, H., Aue, A., Chen, C., Chowdhary, V., Clark, J., Federmann, C., et al. (2018). Achieving human parity on automatic chinese to english news translation. arXiv:1803.05567.
16. He, K., Zhang, X., Ren, S., & Sun, J. (2015). Delving deep into rectifiers: Surpassing human-level performance on imagenet classification. In *Proceedings of the IEEE International Conference on Computer Vision* (pp. 1026–1034).
17. He, K., Zhang, X., Ren, S., & Sun, J. (2016). Deep residual learning for image recognition. In *Proceedings of the IEEE Conference on Computer Vision and Pattern Recognition* (pp. 770–778).
18. He, T., Zhang, Z., Zhang, H., Zhang, Z., Xie, J., & Li, M. (2019). Bag of tricks for image classification with convolutional neural networks. In *Proceedings of the IEEE Conference on Computer Vision and Pattern Recognition* (pp. 558–567).
19. Hochreiter, S., & Schmidhuber, J. (1997). Long short-term memory. *Neural Computation, 9*(8), 1735–1780.
20. Hornik, K., Stinchcombe, M., & White, H. (1990). Universal approximation of an unknown mapping and its derivatives using multilayer feedforward networks. *Neural Networks, 3*(5), 551–560.

21. Huang, C.-Z. A., Vaswani, A., Uszkoreit, J., Simon, I., Hawthorne, C., Shazeer, N., et al. (2019). Music transformer: Generating music with long-term structure. In *International Conference on Learning Representations.*
22. Huang, G., Liu, Z., Van Der Maaten, L., & Weinberger, K. Q. (2017). Densely connected convolutional networks. In *Proceedings of the IEEE Conference on Computer Vision and Pattern Recognition* (pp. 4700–4708).
23. Huang, G.-B., Chen, L., Siew, C. K., et al. (2006). Universal approximation using incremental constructive feedforward networks with random hidden nodes. *IEEE Trans. Neural Networks, 17*(4), 879–892.
24. Hubel, D. H., & Wiesel, T. N. (1968). Receptive fields and functional architecture of monkey striate cortex. *The Journal of Physiology, 195*(1), 215–243.
25. Inoue, H. (2018). Data augmentation by pairing samples for images classification. Preprint. arXiv:1801.02929.
26. Kawaguchi, K. (2016). Deep learning without poor local minima. In *Advances in Neural Information Processing Systems* (pp. 586–594).
27. Kawaguchi, K., & Bengio, Y. (2019). Depth with nonlinearity creates no bad local minima in resnets. *Neural Networks, 118*, 167–174.
28. Kawaguchi, K., Huang, J., & Kaelbling, L. P. (2019). Effect of depth and width on local minima in deep learning. *Neural Computation, 31*(7), 1462–1498.
29. Kingma, D. P., & Ba, J. (2014). Adam: A method for stochastic optimization. Preprint. arXiv:1412.6980.
30. Kobayashi, S. (2018). Contextual augmentation: Data augmentation by words with paradigmatic relations. In *Proceedings of the 2018 Conference of the North American Chapter of the Association for Computational Linguistics: Human Language Technologies, Volume 2 (Short Papers)* (pp. 452–457).
31. Krizhevsky, A., Sutskever, I., & Hinton, G. E. (2012). Imagenet classification with deep convolutional neural networks. In *Advances in Neural Information Processing Systems* (pp. 1097–1105).
32. Lemley, J., Bazrafkan, S., & Corcoran, P. (2017). Smart augmentation learning an optimal data augmentation strategy. *Ieee Access, 5*, 5858–5869.
33. Li, J., Koyamada, S., Ye, Q., Liu, G., Wang, C., Yang, R., et al. (2020). Suphx: Mastering mahjong with deep reinforcement learning. Preprint. arXiv:2003.13590.
34. Liang, S., & Srikant, R. (2019). Why deep neural networks for function approximation? In *5th International Conference on Learning Representations, ICLR 2017.*
35. Lim, S., Kim, I., Kim, T., Kim, C., & Kim, S. (2019). Fast autoaugment. In *Advances in Neural Information Processing Systems* (pp. 6662–6672).
36. Maas, A. L., Hannun, A. Y., & Ng, A. Y. (2013). Rectifier nonlinearities improve neural network acoustic models. In *Proc. ICML* (vol. 30, p. 3).
37. Montufar, G. F., Pascanu, R., Cho, K., & Bengio, Y. (2014). On the number of linear regions of deep neural networks. In *Advances in Neural Information Processing Systems* (pp. 2924–2932).
38. Nair, V., & Hinton, G. E. (2010). Rectified linear units improve restricted boltzmann machines. In *Proceedings of the 27th International Conference on Machine Learning (ICML-10)* (pp. 807–814).
39. Nesterov, Y. (1983). A method for unconstrained convex minimization problem with the rate of convergence o (1/k^ 2). In *Doklady an USSR* (vol. 269, pp. 543–547).
40. Nguyen, Q., & Hein, M. (2017). The loss surface of deep and wide neural networks. In *Proceedings of the 34th International Conference on Machine Learning-Volume 70* (pp. 2603–2612). JMLR.org.
41. Nguyen, Q., & Hein, M. (2018). Optimization landscape and expressivity of deep cnns. In *International Conference on Machine Learning* (pp. 3730–3739).
42. Park, J., & Sandberg, I. W. (1991). Universal approximation using radial-basis-function networks. *Neural Computation, 3*(2), 246–257.

43. Parmar, N., Vaswani, A., Uszkoreit, J., Kaiser, L., Shazeer, N., Ku, A., et al. (2018). Image transformer. In *International Conference on Machine Learning* (pp. 4055–4064).
44. Qian, N. (1999). On the momentum term in gradient descent learning algorithms. *Neural Networks, 12*(1), 145–151.
45. Radford, A., Narasimhan, K., Salimans, T., & Sutskever, I. (2018). Improving language understanding by generative pre-training.
46. Raghu, M., Poole, B., Kleinberg, J., Ganguli, S., & Dickstein, J. S. (2017). On the expressive power of deep neural networks. In *Proceedings of the 34th International Conference on Machine Learning-Volume 70* (pp. 2847–2854). JMLR.org.
47. Ren, Y., Ruan, Y., Tan, X., Qin, T., Zhao, S., Zhao, Z., et al. (2019). Fastspeech: Fast, robust and controllable text to speech. In *Advances in Neural Information Processing Systems* (pp. 3165–3174).
48. Rosenblatt, F. (1958). The perceptron: a probabilistic model for information storage and organization in the brain. *Psychological Review, 65*(6), 386.
49. Shamir, O. (2018). Are resnets provably better than linear predictors? In *Advances in Neural Information Processing Systems* (pp. 507–516).
50. Silver, D., Huang, A., Maddison, C. J., Guez, A., Sifre, L., Van Den Driessche, G., et al. (2016). Mastering the game of go with deep neural networks and tree search. *Nature, 529*(7587), 484.
51. Simonyan, K., & Zisserman, A. (2014). Very deep convolutional networks for large-scale image recognition. Preprint. arXiv:1409.1556.
52. Song, K., Tan, X., Qin, T., Lu, J., & Liu, T.-Y. (2019). Mass: Masked sequence to sequence pre-training for language generation. In *International Conference on Machine Learning* (pp. 5926–5936).
53. Srivastava, N., Hinton, G., Krizhevsky, A., Sutskever, I., & Salakhutdinov, R. (2014). Dropout: a simple way to prevent neural networks from overfitting. *The Journal of Machine Learning Research, 15*(1), 1929–1958.
54. Sutskever, I., Vinyals, O., & Le, Q. V. (2014). Sequence to sequence learning with neural networks. In *Advances in Neural Information Processing Systems* (pp.3104–3112).
55. Szegedy, C., Liu, W., Jia, Y., Sermanet, P., Reed, S., Anguelov, D., et al. (2015). Going deeper with convolutions. In *Proceedings of the IEEE Conference on Computer Vision and Pattern Recognition* (pp. 1–9).
56. Telgarsky, M. (2016). benefits of depth in neural networks. In *Conference on Learning Theory* (pp. 1517–1539).
57. Tibshirani, R. (1996). Regression shrinkage and selection via the lasso. *Journal of the Royal Statistical Society: Series B (Methodological), 58*(1), 267–288.
58. Vaswani, A., Shazeer, N., Parmar, N., Uszkoreit, J., Jones, L., Gomez, A. N., et al. (2017). Attention is all you need. In *Advances in Neural Information Processing Systems* (pp.5998–6008).
59. Wan, L., Zeiler, M., Zhang, S., Le Cun, Y., & Fergus, R. (2013). Regularization of neural networks using dropconnect. In *International Conference on Machine Learning* (pp. 1058–1066).
60. Wang, L.-X., & Mendel, J. M. (1992). Fuzzy basis functions, universal approximation, and orthogonal least-squares learning. *IEEE Transactions on Neural Networks, 3*(5), 807–814.
61. Xie, Z., Wang, S. I., Li, J., Lévy, D., Nie, A., Jurafsky, D., et al. (2019). Data noising as smoothing in neural network language models. In *5th International Conference on Learning Representations, ICLR 2017.*
62. Xiong, W., Droppo, J., Huang, X., Seide, F., Seltzer, M., Stolcke, A., et al. (2016). Achieving human parity in conversational speech recognition. Preprint. arXiv:1610.05256.
63. Zeiler, M. D. (2012). Adadelta: an adaptive learning rate method. Preprint. arXiv:1212.5701.
64. Zhang, H., Cisse, M., Dauphin, Y. N., & Lopez-Paz, D. (2017). mixup: Beyond empirical risk minimization. Preprint. arXiv:1710.09412.

Part II
The Dual Reconstruction Principle

While structure duality among tasks can be interpreted from different perspectives and leveraged in different ways, it was first formally proposed and studied to learn from unlabeled data based on the principle of dual reconstruction. In this part, we will focus on this principle and introduce several dual learning algorithms for different tasks, including machine translation (Chap. 4), image to image translation (Chap. 5), and speech synthesis and recognition (Chap. 6).

Chapter 4
Dual Learning for Machine Translation and Beyond

In this chapter, we focus on neural machine translation and introduce algorithms that leverage unlabeled data based on the principle of dual reconstruction. We first give a brief introduction to machine translation and neural machine translation (NMT), then describe the principle of dual reconstruction, followed by dual semi-supervised algorithms and dual unsupervised algorithms for NMT, and finally introduce several works that leverage the dual reconstruction principle to help other natural language tasks beyond machine translation.

4.1 Introduction to Machine Translation

Machine translation is a sub-field of computational linguistics that investigates the translation of text or speech[1] from one natural language to another using machines. It has a long history: references to the subject can be found as early as the seventeenth century, and a prototype of machine translation came into reality in the 1950s.[2] Since then machine translation has gone through several stages:

1. Rule-Based Machine Translation (RBMT) [27, 28] systems are based on bilingual dictionaries and a set of hand-coded linguistic rules, which are too restrictive for practical applications.
2. Statistical Machine Translation (SMT) [4, 19] translates text using statistical models whose parameters are derived from bilingual text corpora. SMT is a purely data-driven approach and no dictionaries or hand-coded rules are needed.
3. Neural Machine Translation (NMT) [3, 14], the latest and best approach so far, translates text using deep neural networks whose parameters are also derived from bilingual text corpora. Therefore, NMT is also a purely data-driven approach and no dictionaries or hand-coded rules are needed.

[1] We focus on text translation in this chapter.

[2] https://en.wikipedia.org/wiki/Georgetown-IBM_experiment.

T. Qin, *Dual Learning*, https://doi.org/10.1007/978-981-15-8884-6_4

4.1.1 Neural Machine Translation

From the perspective of machine learning, machine translation is a task that transforms a sequence (i.e., a sentence in the source language) into another sequence (i.e., a sentence in the target language). Neural machine translation systems are typically implemented within the encoder-decoder framework: a neural network is used to encode the source sentence, and a neural network is used to decode and generate the target sentence. Such a framework learns a probabilistic mapping $P(y|x)$ from a source language sentence $x = \{x_1, x_2, \ldots, x_{T_x}\}$ to a target language sentence $y = \{y_1, y_2, \ldots, y_{T_y}\}$, in which x_i and y_t are the i-th and t-th words for sentences x and y respectively.

As introduced in Sect. 3.3, the encoder/decoder networks can be recurrent neural networks [3, 36], convolutional neural networks [11], and Transformer [38]. For simplicity, we take recurrent neural networks (RNNs) as an example to introduce how neural machine translation works.

The encoder of NMT reads the source sentence x and generates T_x hidden states by an RNN:

$$h_i = f(h_{i-1}, x_i) \tag{4.1}$$

in which h_i is the hidden state at position i, and function f is the recurrent unit such as Long Short-Term Memory (LSTM) unit [36] or Gated Recurrent Unit (GRU) [7]. Afterwards, the decoder of NMT computes the conditional probability of each target word y_t given its proceeding words $y_{<t}$, as well as the source sentence, i.e., $P(y_t|y_{<t}, x)$, which is then used to specify $P(y|x)$ according to the probability chain rule. $P(y_t|y_{<t}, x)$ is given as:

$$P(y_t|y_{<t}, x) \propto \exp(y_t; s_t, c_t) \tag{4.2}$$

$$s_t = g(s_{t-1}, y_{t-1}, c_t) \tag{4.3}$$

$$c_t = q(s_{t-1}, h_1, \cdots, h_{T_x}) \tag{4.4}$$

where s_t is the decoder RNN hidden state at time t, similarly computed by an LSTM or GRU, and c_t denotes the contextual information in generating word y_t according to different encoder hidden states. c_t can be a 'global' signal summarizing sentence x [7, 36], e.g., $c_1 = \cdots = c_{T_y} = h_{T_x}$, or 'local' signal implemented by an attention mechanism [3], e.g.,

$$c_t = \sum_{i=1}^{T_x} \alpha_i h_i,$$

$$\alpha_i = \frac{\exp\{a(h_i, s_{t-1})\}}{\sum_j \exp\{a(h_j, s_{t-1})\}},$$

where $a(\cdot, \cdot)$ is a feed-forward neural network.

We denote all the parameters to be optimized in the neural network as θ and denote $\mathcal{D}$ as the dataset that contains source-target sentence pairs for training. Then the training of NMT is to seek the optimal parameters θ^* that maximize data likelihood:

$$\theta^* = \arg\max_{\theta} \frac{1}{|\mathcal{D}|} \sum_{(x,y)\in\mathcal{D}} \sum_{t=1}^{T_y} \log P(y_t|y_{<t}, x; \theta) \tag{4.5}$$

4.1.2 Back Translation

Both SMT and NMT rely on large parallel/bilingual corpora which contains paired sentences in both the source and target language. However, bitext is limited and costly to collect, while large scale of monolingual data is easy to collect at low cost. Monolingual data has been widely used in SMT to train language models and improve the fluency of translations [19].

In neural machine translation, different approaches have been proposed to improve models with monolingual data before dual learning, including language model fusion [12, 13] and back translation [31]. Since dual learning is related to back translation, we introduce back translation in this sub section.

The basic idea of back translation is simple. Suppose we are working on the translation from the source language X to the target language Y. The goal of back translation is to improve the translation using monolingual data of the target language Y. It first trains a reverse translation model that translates a sentence y in language Y to a sentence x' in language X. We denote the set of such synthetic sentence pairs as $\mathcal{D}_S$. It then trains the $X \to Y$ translation model using both the human-labeled sentence pairs $\mathcal{D}_H$ and the synthetic ones $\mathcal{D}_S$:

$$\theta^* = \arg\max_{\theta} \sum_{(x,y)\in\mathcal{D}_H} P(y|x;\theta) + \sum_{(x',y)\in\mathcal{D}_S} P(y|x';\theta). \tag{4.6}$$

The effectiveness of back translation has been verified in both small-scale studies [31] and large-scale studies [10].

4.2 The Principle of Dual Reconstruction

Structure duality has been leveraged to enhance neural machine translation with unlabeled data in different settings, including semi-supervised learning [15], unsupervised learning, and multi-agent learning [39], all of which share the same reconstruction principle. We introduce the principle in this section.

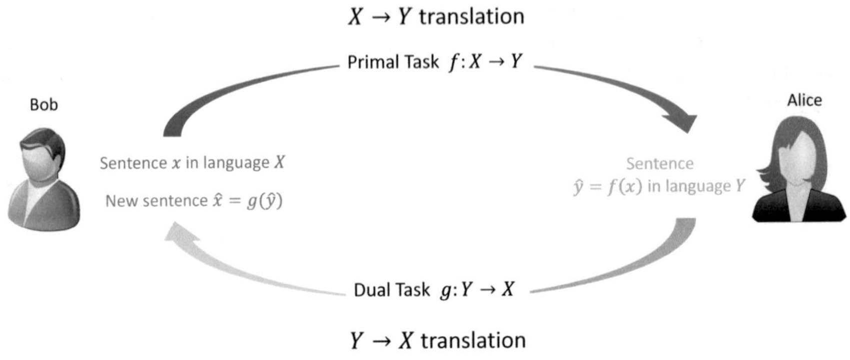

Fig. 4.1 The reconstruction principle

Noticing that machine translation can (always) happen in dual directions (i.e., from X to Y and from Y to X), dual learning formulates bidirectional translations as a two-agent communication game, as shown in Fig. 4.1.

In the game, two agents Bob, who only speaks language X, and Alice, who only speaks language Y, want to communicate with each other. They employ two translation models to enable this cross-lingual communication. When Bob wants to talk to Alice,

1. He sends the message x in language X through a translation channel with the forward model f to Alice.
2. Alice receives the message $\hat{y} = f(x)$ in language Y.
3. To check whether $\hat{y}$ is exactly what Bob wants to say, Alice further sends $\hat{y}$ to Bob through another translation channel with the backward model g.
4. Bob receives a new message $\hat{x} = g(\hat{y})$ in language X, and checks with the original message x.

Clearly, if the two translation models work well, x should be able to be reconstructed from its translation $\hat{y}$, and therefore $\hat{x}$ should be semantically similar to or exactly same as x. If $\hat{x}$ is far from x, we know that something is wrong with (at least one of) the two translation/communication channels. That is, through such a communication game, we get feedback signals about the quality of the two translation models, which can then be used to update and improve the two models. Note that here we do not need to know the correct translation of x.

Similarly, when Alice wants to talk to Bob, she can start from a message y in language Y, and sends through the channel with the translation model g; Bob receives $\hat{x} = g(y)$ in language X and sends it through the channel with the translation model f; Alice receives the message $\hat{y} = g(\hat{x})$ in language Y and compares it with her original message y.

To summarize, if the two models of two dual tasks are perfect, the original input should be reconstructed by the two models:

$$x = g(f(x))$$

and

$$y = f(g(y)).$$

In other words, we can minimize the reconstruction error

$$\min_{f,g} \Delta(x, g(f(x))), \tag{4.7}$$

$$\min_{f,g} \Delta(y, f(g(y))) \tag{4.8}$$

to improve the two models. This is so called the *deterministic dual reconstruction principle*.

In machine translation, a neural network model usually translates a source sentence into multiple candidate sentences in the target language with different probabilities. Therefore, instead of minimize the reconstruction error, we maximize the reconstruction probability, which is equivalent to minimizing the negative log likelihood of dual reconstruction:

$$\min_{f,g} \ell(x; f, g) = \min_{f,g} - \log P(x|f(x); g); \tag{4.9}$$

$$\min_{f,g} \ell(y; f, g) = \min_{f,g} - \log P(y|g(y); f); \tag{4.10}$$

which is called the *probabilistic dual reconstruction principle*.

While we take machine translation as example to elaborate the principle here, it can be applied to many other problems. More examples will be discussed in Chaps. 5 and 6.

4.3 Dual Semi-supervised Learning

In this section, we introduce the work of dual learning for machine translation in the semi-supervised setting [15], i.e., using both labeled data (human-labeled parallel sentences) and unlabeled data (monolingual sentences in both the source language and target language). We call the algorithm *DualNMT*.

Consider two monolingual corpora $\mathcal{M}_X$ and $\mathcal{M}_Y$ which contain sentences from language X and Y respectively. These two corpora are not necessarily aligned with each other, and they may even have no topical relationship with each other at all. Suppose we have two (weak) translation models that can translate sentences from X

to Y and verse visa. Our goal is to improve the accuracy of the two models by using monolingual corpora instead of parallel corpora. The basic idea of DualNMT is to leverage the dual structure of the two translation tasks and maximize the likelihood of dual reconstruction. As described in previous section, starting from a sentence in any monolingual data, DualNMT first translates it forward to the other language and then further translates backward to the original language. By evaluating this two-hop translation results, DualNMT gets a sense about the quality of the two translation models, and then improves them accordingly.

Algorithm 1 The DualNMT algorithm

1: **Input**: Monolingual corpora $\mathcal{M}_X$ and $\mathcal{M}_Y$, initial primal translation model θ_{XY} and dual translation model θ_{YX}, language models $P_X()$ and $P_Y()$, hyper-parameter α, beam search size K, learning rates $\gamma_{1,t}$, $\gamma_{2,t}$.
2: Set $t = 0$.
3: **repeat**
4: $t = t + 1$.
5: Sample two sentence s_X and s_Y from $\mathcal{M}_X$ and $\mathcal{D}_Y$ respectively.
6: Set $s = s_X$. ▷ *The game beginning from X.*
7: Generate K sentences $s_{mid,1}, \cdots, s_{mid,K}$ using beam search according to translation model θ_{XY}.
8: **for** $k = 1, \ldots, K$ **do**
9: Set the language-model reward for the k-th sampled sentence as $r_{1,k} = P_Y(s_{mid,k})$.
10: Set the reconstruction reward for the k-th sampled sentence as

$$r_{2,k} = \log P(s|s_{mid,k}; \theta_{YX}).$$

11: Set the total reward of the k-th sample as $r_k = \alpha r_{1,k} + (1 - \alpha) r_{2,k}$.
12: **end for**
13: Compute the stochastic gradient of θ_{XY}:

$$\nabla_{\theta_{XY}} \hat{E}[r] = \frac{1}{K} \sum_{k=1}^{K} [r_k \nabla_{\theta_{XY}} \log P(s_{mid,k}|s; \theta_{XY})].$$

14: Compute the stochastic gradient of θ_{YX}:

$$\nabla_{\theta_{YX}} \hat{E}[r] = \frac{1}{K} \sum_{k=1}^{K} [(1 - \alpha) \nabla_{\theta_{YX}} \log P(s|s_{mid,k}; \theta_{YX})].$$

15: Model updates:

$$\theta_{XY} \leftarrow \theta_{XY} + \gamma_{1,t} \nabla_{\theta_{XY}} \hat{E}[r], \theta_{YX} \leftarrow \theta_{YX} + \gamma_{2,t} \nabla_{\theta_{YX}} \hat{E}[r].$$

16: Set $s = s_Y$. ▷ *The game beginning from Y.*
17: Go through line 6 to line 15 symmetrically.
18: **until** convergence

In DualNMT, it is assumed that the two well-trained language models $P_X()$ and $P_Y()$ are pre-given. A language model [25, 35] takes a partial or complete sentence as input and outputs a real value to indicate how likely the sentence is a natural sentence. Note that only monolingual data is required to train a language model. For example, one can simply use the monolingual data $\mathcal{M}_X$ and $\mathcal{M}_Y$ to train $P_X()$ and $P_Y()$.

Let θ_{XY} and θ_{YX} denote two neural translation models parameterized by θ_{XY} and θ_{YX} respectively. For a game beginning with sentence s in $\mathcal{M}_X$, denote s_{mid} as the middle translation output. This middle step has an immediate feedback signal $r_1 = P_Y(s_{mid})$, indicating how likely s_{mid} is a natural sentence in language B. Given the middle translation output s_{mid}, the log probability of s reconstructed from s_{mid} is used as the reward of the reconstruction. Mathematically, reward $r_2 = \log P(s|s_{mid}; \theta_{YX})$.

DualNMT simply adopts a linear combination of the LM reward and reconstruction reward as the total reward, e.g., $r = \alpha r_1 + (1 - \alpha)r_2$, where α is a hyper-parameter. As the reward of the game can be considered as a function of s, s_{mid} and translation models θ_{XY} and θ_{YX}, DualNMT optimizes the parameters of the two translation models using policy gradient methods, which are designed for reinforcement learning [37].

DualNMT samples s_{mid} according to the translation model $P(.|s; \theta_{XY})$. It then computes the gradient of the expected reward $E[r]$ with respect to parameters θ_{XY} and θ_{YX}. According to the policy gradient theorem [37], it is easy to verify that

$$\nabla_{\theta_{YX}} E[r] = E[(1-\alpha)\nabla_{\theta_{YX}} \log P(s|s_{mid}; \theta_{YX})] \tag{4.11}$$

$$\nabla_{\theta_{XY}} E[r] = E[r\nabla_{\theta_{XY}} \log P(s_{mid}|s; \theta_{XY})] \tag{4.12}$$

in which the expectation is taken over s_{mid}.

Based on Eqs. (4.11) and (4.12), one can adopt any sampling approach to estimate the expected gradient. Considering that random sampling brings very large variance and sometimes unreasonable results in machine translation [29, 30, 36], He et al. [15] use beam search [36] to obtain more meaningful results (more reasonable middle translation outputs) for gradient computation, i.e., greedily generating top-K high-probability middle translation outputs, and use the averaged value on the beam search results to approximate the true gradient. If the game begins with sentence s in $\mathcal{M}_Y$, the gradient can be computed similarly.

The game can be repeated for many rounds. In each round, one sentence is sampled from $\mathcal{M}_X$ and one from $\mathcal{M}_Y$, and the two models are updated according to the game beginning with the two sentences respectively. The details of DualNMT are given in Algorithm 1. Since DualNMT takes two models θ_{XY} and θ_{YX} as inputs, which are pre-trained from labeled bitext,[3] it is a semi-supervised algorithm.

As pointed out in [15], the idea of dual reconstruction and so the DualNMT algorithm are not restricted to two tasks. Actually, the basic idea is to form a closed

[3] 1+ million bilingual sentence pairs were used to pre-train the two models in [15].

loop so that one can get feedback signals by computing the reconstruction error between the original input data with the final output data. Therefore, if more than two associated tasks can form a closed loop, one can apply the same idea to improve the model in each task from unlabeled data. For example, for an English sentence x, one can first translate it to a Chinese sentence y, then translate y to a French sentence z, and finally translate z back to an English sentence x'. The reconstruction error between x and x' can indicate the effectiveness of the three translation models in the loop, and one can apply policy gradient methods to update and improve these models based on the feedback signals during the loop. This generalized dual learning is named as *close-loop learning* [15].

4.3.1 Zero-Shot Dual Machine Translation

The above DualNMT algorithm needs some paired data to train the two weak translation models, which are then used to kick off the dual learning process. Sestorain et al. [32] study dual learning for zero-shot machine translation, in which there is no paired data available. That is, they consider the problem of unsupervised machine translation, and their approach is built on multilingual NMT architecture.

While the approach proposed in [32] works for three and more than three languages, for simplicity we use three languages to introduce it. Let X, Y and Z denote the three languages. Let the pairs X-Z and Y-Z be the language pairs with parallel data and X-Y be the target language pair without parallel data. It is assumed that (1) sufficient monolingual data $\mathcal{M}_X$ and $\mathcal{M}_Y$ for languages X and Y are available, and (2) sufficient bilingual data $\mathcal{B}_{XZ}$ and $\mathcal{B}_{YZ}$ for language pairs X-Z and Y-Z are available.

Before start zero-shot dual learning, we need to first pre-train a single multilingual NMT model with parameters θ using bilingual data $\mathcal{B}_{XZ}$ and $\mathcal{B}_{YZ}$. While this model does not train on any X-Y sentence pairs, it can do zero-shot translation for X-Y, i.e., translation from X to Y and from Y to X, although the quality might be poor. That is, this approach kicks off the dual learning process with a multilingual translation model, while DualNMT with weak models trained on a small amount of paired data. Furthermore, we need to pre-train language models $P_X()$ and $P_Y()$ for X and Y using monolingual data $\mathcal{M}_X$ and $\mathcal{M}_Y$.

Similar to [15], Sestorain et al. [32] employ a REINFORCE-like algorithm for zero-shot dual learning.

1. Sample a sentence x from $\mathcal{M}_X$ and sample a corresponding translation from the multilingual NMT model $\hat{y} \sim P_\theta(\cdot|x)$.
2. Compute the language model score $r_1 = \log P_Y(\hat{y})$.
3. Compute the reconstruction reward $r_2 = \log P_\theta(x|\hat{y})$.
4. Compute the total reward for $\hat{y}$ as $R = \alpha r_1 + (1 - \alpha) r_2$, where α is a hyper-parameter.
5. Update θ using the REINFORCE method with the reward R.

This update process can also be started with a sentence sampled from $\mathcal{M}_Y$. The whole training will iterate between X language sentences and Y language sentences until certain termination criterion is satisfied.

4.4 Dual Unsupervised Learning

As introduced in the previous section, DualNMT requires bilingual data to pre-train the two models. While it is not very difficult to collect bilingual data for top language pairs, it is indeed difficult for rare language pairs. Note that there are more than 7000 languages[4] that are spoken today in the world, and most of them are rare languages. It is of great importance to enable machine translation without any bilingual data, so as to protect those rare languages and also help the native speakers to better communicate with the majority of the world. Multiple methods have been proposed for unsupervised neural machine translation, in which structure duality plays an important role. We introduce two representative and contemporary works [2, 20]. We first describe the basic ideas underlying the two works, and then the system architectures and training algorithms.

4.4.1 *Basic Ideas*

Although the system architectures and training algorithms are different in [2] and [20], they share the same basic ideas.

Both the works adopt the encoder-decoder framework as standard neural machine translation. They both use an encoder to build a common latent space shared by two languages (i.e., the source language and the target language). Let θ_{en}^A denote the encoder and θ_{de}^A the decoder for language A, and similarly define θ_{en}^B and θ_{de}^B. The encoders and decoders are trained using monolingual data based on two principles: denoising auto-encoding and dual reconstruction.

Denoising auto-encoding can also be viewed as denoising reconstruction: a sentence of a language should be reconstructed from its corrupted version using the encoder and decoder of that language. Mathematically, we minimize the denoising reconstruction error

$$\ell_{dae}(\theta_{en}^l, \theta_{de}^l) = \frac{1}{|\mathcal{M}_l|} \sum_{x \in \mathcal{M}_l} \Delta(x, \theta_{de}^l(\theta_{en}^l(c(x)))),$$

or the negative log likelihood of denoising reconstruction

$$\ell_{dae}(\theta_{en}^l, \theta_{de}^l) = -\frac{1}{|\mathcal{M}_l|} \sum_{x \in \mathcal{M}_l} \log P(x|\theta_{de}^l(\theta_{en}^l(c(x)))),$$

[4] https://www.ethnologue.com/guides/how-many-languages.

where $l \in \{A, B\}$ denotes the source or target language, $\mathcal{M}_l$ denotes the monolingual corpus for language l, $c(x)$ denotes a corrupted version of sentence x, and $\theta^l_{de}(\theta^l_{en}(c(x)))$ denotes the reconstructed sentence from the corrupted version $c(x)$. Standard auto-encoding is not used considering that it may learn to simply copy the words one by one from the input sentence to the output.

Dual reconstruction aims to reconstruct any sentence in one language from its noisy translation[5] from the other language. Mathematically, we minimize the dual reconstruction error

$$\begin{aligned}\ell_{dual}(\theta^l_{en}, \theta^l_{de}) = \tfrac{1}{|\mathcal{M}_A|} \textstyle\sum_{x\in\mathcal{M}_A} \Delta(x, \theta^A_{de}(\theta^B_{en}(c(\hat{y})))) \\ + \tfrac{1}{|\mathcal{M}_B|} \textstyle\sum_{x\in\mathcal{M}_B} \Delta(x, \theta^B_{de}(\theta^A_{en}(c(\hat{y})))),\end{aligned}$$

or the negative log likelihood of denoising reconstruction

$$\begin{aligned}\ell_{dual}(\theta^l_{en}, \theta^l_{de}) = -\tfrac{1}{|\mathcal{M}_A|} \textstyle\sum_{x\in\mathcal{M}_A} \log P(x|\theta^A_{de}(\theta^B_{en}(c(\hat{y})))) \\ -\tfrac{1}{|\mathcal{M}_B|} \textstyle\sum_{x\in\mathcal{M}_B} \log P(x|\theta^B_{de}(\theta^A_{en}(c(\hat{y})))),\end{aligned}$$

where $\hat{y} = \theta^B_{de}(\theta^A_{en}(x))$ is the translation of a sentence x in language A, $\hat{y} = \theta^A_{de}(\theta^B_{en}(x))$ is the translation of a sentence x in language B, and $\theta^{l'}_{de}(\theta^l_{en}(c(\hat{y})))$ is the reconstructed sentence from the corrupted translation $c(\hat{y})$.

In the following two sub sections, we focus on unsupervised neural machine translation [2, 20]. The same principles have also been studied in unsupervised phrase-based machine translation [21].

4.4.2 System Architectures and Training Algorithms

Artetxe et al. [2] use a shared encoder for the two languages but different decoders, as shown in Fig. 4.2. That is, we have $\theta^A_{en} = \theta^B_{en}$ but $\theta^A_{de} \neq \theta^B_{de}$. Note that although the encoder is shared, the vocabularies are not shared between the two languages.

In contrast, Lample et al. [20] share both the encoder and decoder for the two languages, i.e., $\theta^A_{en} = \theta^B_{en}$ and $\theta^A_{de} = \theta^B_{de}$. They simply use two embedding vectors to differentiate the two languages.

Artetxe et al. [2] first pre-train cross-lingual embeddings in the shared encoder. They use the monolingual corpora $\mathcal{M}_A$ and $\mathcal{M}_B$ to independently train the embeddings for each language using the skip-gram model [26]. Then these embeddings are mapped to a shared space using the method[6] proposed in [1]. After pre-training,

[5]Since both translation models are not perfect, we can only get a noisy translation for a sentence.

[6]The public implementation can be found at https://github.com/artetxem/vecmap.

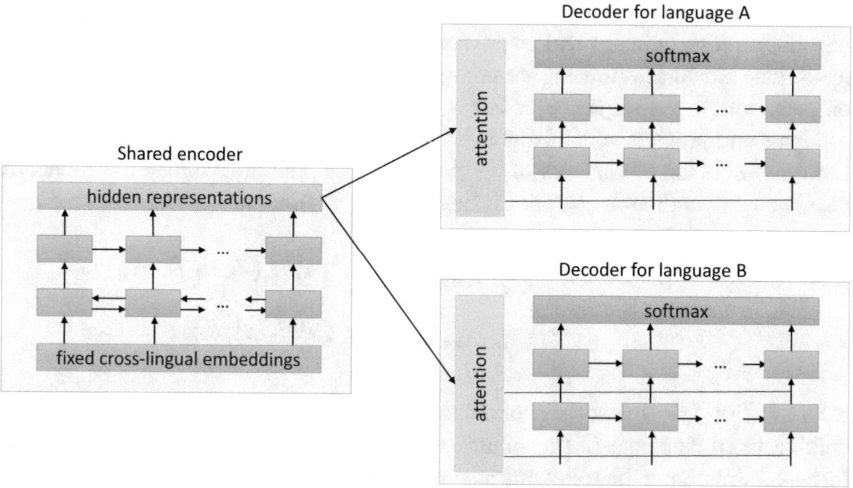

Fig. 4.2 Neural network architecture for unsupervised machine translation in [2]

the embeddings are fixed during unsupervised training. This is different from most NMT systems, in which word embeddings are randomly initialized and then updated during training.

Second, with pre-trained embeddings, the shared encoder and two decoders are jointly trained through denoising auto-encoding and dual reconstruction. To corrupt a sentence for denoising auto-encoding, Artetxe et al. randomly swap two consecutive words, e.g, for a sequence of N words/tokens, they make $N/2$ random swaps of this kind. Doing so, the system is forced to discover the internal structure of the languages so as to recover the correct word order. Furthermore, Artetxe et al. argue that by discouraging the system to rely too much on the word order of the input sequence, the trained system can better handle the actual word order divergences across languages. During training, the objectives of denoising auto-encoding and dual reconstruction are used separately from mini-batch to mini-batch. That is, one mini-batch is trained through denoising auto-encoding for language A, another one for language B, one mini-batch is trained through dual reconstruction from A to B and then A, and another one from B to A and then B.

Artetxe et al. further show that their proposed system can be enhanced and trained in a semi-supervised manner when a small parallel corpus is provided (see Table 1 in [2]).

While the system architecture in [20] is simpler than that in [2], its training process is more complex. Lample et al. [20] starts with a simple word-by-word translation model, which is actually a parallel dictionary learned from monolingual data in an unsupervised way [8]. Then, at each iteration, the encoder and decoder are trained by minimizing the loss of denoising auto-encoding and dual reconstruction from a noisy/corrupted input sentence, which is obtained by dropping and swapping words from the original sentence.

In addition to denoising auto-encoding and dual reconstruction, Lample et al. introduce a discriminator and an adversarial loss to ensure that the encoder indeed maps two languages into the common latent space. The discriminator takes the outputs, which is a sequence of latent vectors in the latent space, of the encoder as inputs and predicts which language the original input sentence of the encoder is. Let θ_D denote the parameters of the discriminator. The discriminator is trained to minimize the prediction error or the negative log likelihood:

$$\begin{aligned} l_D(\theta_D) = & -\frac{1}{|\mathcal{M}_A|} \sum_{x \in \mathcal{M}_X} \log P(A|\theta_{en}^A(x); \theta_D) \\ & -\frac{1}{|\mathcal{M}_Y|} \sum_{x \in \mathcal{M}_B} \log P(B|\theta_{en}^B(x); \theta_D), \end{aligned}$$

where $P(l|\theta_{en}^l(x); \theta_D)$ is the probability that the discriminator predicts sentence x coming from language l. The shared encoder is trained to fool the discriminator through minimizing the following adversarial loss:

$$\begin{aligned} \ell_{adv}(\theta_{en}) = & -\frac{1}{|\mathcal{M}_A|} \sum_{x \in \mathcal{M}_A} \log P(B|\theta_{en}^A(x); \theta_D) \\ & -\frac{1}{|\mathcal{M}_B|} \sum_{x \in \mathcal{M}_B} \log P(A|\theta_{en}^B(x); \theta_D), \end{aligned}$$

In each training iteration/min-batch, the translation model is updated to minimize the following total loss,

$$\begin{aligned} \ell(\theta_{en}^l, \theta_{de}^l) = & \lambda_{dae} \ell_{dae}(\theta_{en}^l, \theta_{de}^l) \\ & + \lambda_{dual} \ell_{dual}(\theta_{en}^l, \theta_{de}^l) \\ & + \lambda_{adv} \ell_{adv}(\theta_{en}^l), \end{aligned}$$

where λ_{dae}, λ_{dual} and λ_{adv} are hyper-parameters to trade off the denoising auto-encoding, dual reconstruction, and adversarial loss; in parallel, the discriminator is updated to minimize its prediction error. Figure 4.3 illustrates the system architecture together with the loss for each component.

Lample et al. [20] have some observations. (1) All the three components (the denoising auto-encoding, dual reconstruction, and adversarial loss) are helpful and the best performance is achieved with all of them. (2) The dual reconstruction loss is the most critical among the three components: without this loss, the BLEU score can drop up to 20 points (see Table 4 in [20]), which demonstrates the importance of dual reconstruction.

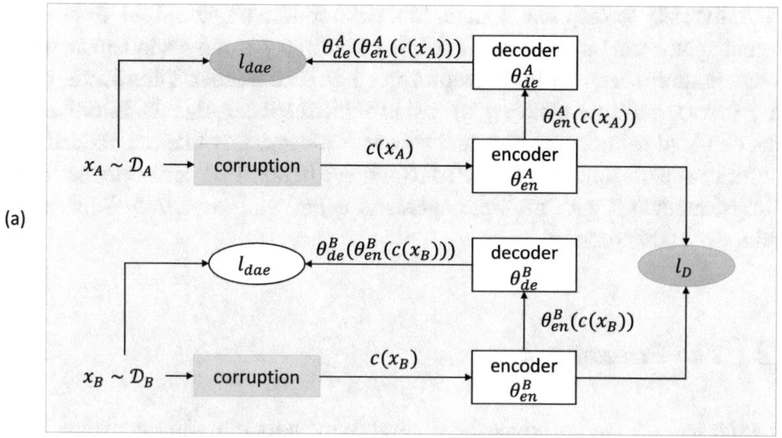

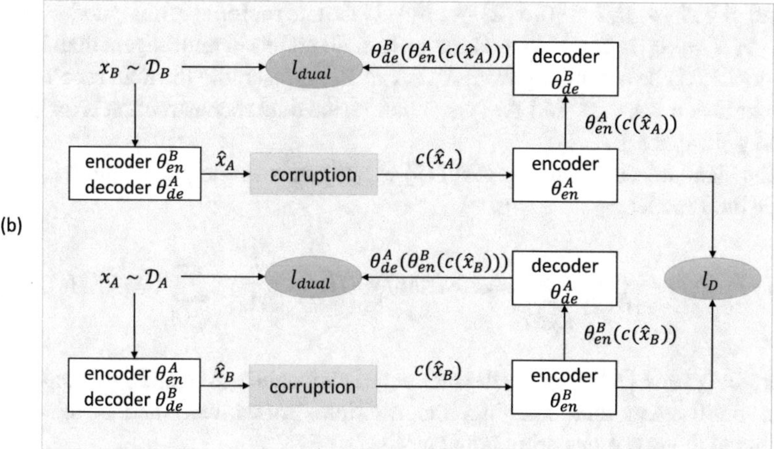

Fig. 4.3 Illustration of the system architecture and training loss in [20]. (**a**) Denoising auto-encoder and its training loss. (**b**) Dual reconstruction and its training loss

4.5 Multi-Agent Dual Learning

In the two-agent cross-lingual communication game/system described in Sect. 4.2, g and f can be viewed as an evaluator for each other. g is used to evaluate the quality of $\hat{y}$ generated by f and return the feedback signal $\Delta_X(x, g(\hat{y}))$ back to f, and vice versa. The quality of such evaluation plays a central role to improve the primal and dual models. Those dual learning works introduced so far in this chapter only leverage one agent g to evaluate and provide feedback signals to the model f in the other direction. Inspired by ensemble learning [9], Wang et al. [39] introduce multiple agents into the learning system to further exploit the potential of dual learning. The agents in the same direction have similar capability and certain level of diversity to map one domain to the other domain, i.e., $\mathcal{X} \to \mathcal{Y}$ or $\mathcal{Y} \to \mathcal{X}$. Different agents can be obtained by training multiple f's and g's independently with different random seeds for initialization and data access order. Then for the output of each f (or g), multiple g's (or f's) will provide feedback signals. Intuitively, more agents can lead to more reliable and robust feedback, just like the majority voting of multiple experts, and it is expected to achieve better final performance. This dual learning framework with multiple agents is called *multi-agent dual learning*. We introduce it in this section.

4.5.1 The Framework

In this sub section, we introduce the general framework of multi-agent dual learning. We will discuss how to apply it to machine translation in next sub section.

Let $f_i : \mathcal{X} \to \mathcal{Y}, i \in \{0, 1, 2 \cdots, N-1\}$ denote multiple primal models and g_i : $\mathcal{Y} \to \mathcal{X}, i = \{0, 1, 2, \cdots, N-1\}$ multiple dual models in multi-agent dual learning. Let $\Delta_X(x, x')$ be a mapping from $\mathcal{X} \times \mathcal{X}$ to $\mathbb{R}$, representing the dual reconstruction error between x and x', and $\Delta_Y(y, y')$ denote the dual reconstruction error between y and y' in space $\mathcal{Y}$.

The standard dual learning loss [15] is built on a single primal model f_0 and a single dual model g_0

$$\ell_{dual}(f_0, g_0) = \frac{1}{|\mathcal{M}_X|} \sum_{x \in \mathcal{M}_X} \Delta_X(x, g_0(f_0(x))) + \frac{1}{|\mathcal{M}_Y|} \sum_{y \in \mathcal{M}_Y} \Delta_Y(y, f_0(g_0(y))),$$

where $|\mathcal{M}_X|$ and $|\mathcal{M}_Y|$ denote the number of elements in $\mathcal{M}_X$ and $\mathcal{M}_Y$ respectively.

In multi-agent dual learning, the multiple primal/dual models are linearly combined to get a super primal/dual model:

$$F_\alpha = \sum_{i=0}^{N-1} \alpha_i f_i, \quad \text{s.t.} \sum_{i=0}^{N-1} \alpha_i = 1, \tag{4.13}$$

$$G_\beta = \sum_{j=0}^{N-1} \beta_j g_j, \quad \text{s.t.} \sum_{j=0}^{N-1} \beta_j = 1, \tag{4.14}$$

where $1 \geq \alpha_i \geq 0$ and $1 \geq \beta_i \geq 0$ are weights for linear combination. Then the duality feedback signal is built upon F_α and G_β. Following the basic framework of dual learning [15], for any $x \in \mathcal{X}$, all agents first cooperate to generate a $\hat{y} \in \mathcal{Y}$ by $\hat{y} = F_\alpha(x)$, jointly reconstruct the $\hat{x} \in \mathcal{X}$ through $\hat{x} = G_\beta(\hat{y})$, and then compute the reconstruction error $\Delta_X(x, \hat{x})$. The reconstruction error between $y \in \mathcal{Y}$ and $\hat{y} = F_\alpha(G_\beta(y))$ is similarly calculated. The out-coming dual reconstruction loss for multi-agent dual learning becomes

$$\ell_{dual}(F_\alpha, G_\beta) = \frac{1}{|\mathcal{M}_X|} \sum_{x \in \mathcal{M}_X} \Delta_X(x, G_\beta(F_\alpha(x))) + \frac{1}{|\mathcal{M}_Y|} \sum_{y \in \mathcal{M}_Y} \Delta_Y(y, F_\alpha(G_\beta(y))). \quad (4.15)$$

While in principle the $2N$ primal and dual models can be trained jointly by minimizing the above dual reconstruction error, it will lead to huge computational cost. Wang et al. [39] suggest the following training procedure:

1. Pre-train each f_i and g_i separately following standard training procedure of neural machine translation, e.g., minimizing $-\sum_{(x,y)} \log P(y|x; f_i)$ via stochastic gradient descent over paired data (x, y) [3], or using unsupervised learning techniques to obtain like unsupervised NMT [1, 20].
2. Fix $2(N-1)$ models f_i and g_i for $i \in \{1, 2, \cdots, N-1\}$, and only train f_0 and g_0 to minimize the dual reconstruction error $\ell()$ defined in Eq. (4.15). That is, to reduce computational cost, only one primal model and one dual model are trained in multi-agent dual learning.

To ensure the performance of multi-agent dual learning, one should introduce diversity among those multiple agents, which can be implemented in many ways. For example, one can adopt different neural network structures for those models, initialize them with different random seeds, feed them with different data access/shuffling orders in pre-training, or pre-train them with different data and settings such as supervised training with paired data, unsupervised training with unpaired data, and semi-supervised training with both paired and unpaired data.

4.5.2 Extensions and Comparisons

We discuss possible extensions of multi-agent dual learning and compare it with other related learning paradigms.

The training objective in Eq. (4.15) is about dual reconstruction. Other training objectives could also be included. For example, in machine translation, if bilingual/paired data is available, the maximum likelihood loss can be included to guide the training (see next sub section); in image-to-image translation, the GAN loss could also be included to enforce the generated images into the correct categories.

While there exist many works using multiple agents to boost model performance, none of them has touched structure duality. We take the primal task $\mathcal{X} \to \mathcal{Y}$ as an example to compare previous works with multi-agent dual learning.

Ensemble learning [43] is a straightforward way to combine multiple models during inference. To predict the label of $x \in \mathcal{X}$, all agents vote together and the final label would be

$$\arg\min_{y \in \mathcal{Y}} \sum_{i=0}^{N-1} \alpha_i \ell(f_i(x), y),$$

where ℓ is the loss function over space $\mathcal{Y} \times \mathcal{Y}$. The α_i's can be simply set as $1/N$ or adaptively set according to the quality of each agent. There are several differences between ensemble learning and multi-agent dual learning. First, ensemble learning does not use multiple agents in training as multi-agent dual learning does. Second, multi-agent dual learning uses only one model f_0 in inference, which is more efficient than ensemble learning which uses multiple models. Third, structure duality is not considered in conventional ensemble learning.

Knowledge distillation with multiple teachers [16, 18] consists of two steps: First, all teachers f_i's generate soft labels for $x \in \mathcal{X}$, e.g.,

$$\hat{y} = \arg\min_{y \in \mathcal{Y}} \sum_{i=0}^{N-1} \alpha_i \ell(f_i(x), y).$$

Second, the generated pairs $(x, \hat{y})$'s are used together to train a student model. The main difference is that in knowledge distillation, each $(x, \hat{y})$ is regarded as labeled data without evaluating the quality of $\hat{y}$ and whether it is good enough to train the student model. In contrast, multi-agent dual learning leverages structure duality to build a feedback loop so as to evaluate the quality of a generated pseudo pair $(x, \hat{y})$.

4.5.3 Multi-Agent Dual Machine Translation

As the focus of this chapter is machine translation, in this sub section, we show how multi-agent dual learning can be implemented for machine translation.

Denote the parameters of f_0 and g_0 as θ_0^f and θ_0^g respectively. Following the probabilistic dual reconstruction principle in Eq. (4.9), the Δ_X and Δ_Y are specified as negative log-likelihood: for any $x \in \mathcal{X}$,

$$\begin{aligned}
&\Delta_X(x, G_\beta(F_\alpha(x))) = -\log P(x|F_\alpha(x); G_\beta) \\
&= -\log \sum_{\hat{y} \in \mathcal{Y}} P(F_\alpha(x) = \hat{y}|x; F_\alpha, G_\beta) P(G_\beta(\hat{y}) = x|x, F_\alpha(x) = \hat{y}; F_\alpha, G_\beta) \\
&= -\log \sum_{\hat{y} \in \mathcal{Y}} P(F_\alpha(x) = \hat{y}|x; F_\alpha) P(G_\beta(\hat{y}) = x|\hat{y}; G_\beta) \\
&\overset{\text{briefly}}{=} -\log \sum_{\hat{y} \in \mathcal{Y}} P(\hat{y}|x; F_\alpha) P(x|\hat{y}; G_\beta).
\end{aligned}$$

Similarly, for any $y \in \mathcal{Y}$, we have

$$\Delta_Y(y, F_\alpha(G_\beta(y))) = -\log \sum_{\hat{x} \in \mathcal{X}} P(\hat{x}|y; G_\beta) P(y|\hat{x}; F_\alpha).$$

To simplify the above optimization problem, one can minimize the following upper bounds of Δ_X and Δ_Y:

$$\bar{\Delta}_x(x, G_\beta(F_\alpha(x))) = -\sum_{\hat{y}\in\mathcal{Y}} P(\hat{y}|x; F_\alpha) \log P(x|\hat{y}; G_\beta) \geq \Delta_X(x, G_\beta(F_\alpha(x)));$$

$$\bar{\Delta}_y(y, F_\alpha(G_\beta(y))) = -\sum_{\hat{x}\in\mathcal{X}} P(\hat{x}|y; G_\beta) \log P(y|\hat{x}; F_\alpha) \geq \Delta_Y(y, F_\alpha(G_\beta(y))).$$

The two $\geq$ hold due to Jensen's inequality. Then one turns to minimize

$$\begin{aligned}\tilde{\ell}_{dual}(\mathcal{M}_X, \mathcal{M}_Y; F_\alpha, G_\beta) = &\tfrac{1}{|\mathcal{M}_X|} \textstyle\sum_{x\in\mathcal{M}_X} \bar{\Delta}_x(x, G_\beta(F_\alpha(x))) \\ &+ \tfrac{1}{|\mathcal{M}_Y|} \textstyle\sum_{y\in\mathcal{M}_Y} \bar{\Delta}_y(y, F_\alpha(G_\beta(y))).\end{aligned}$$

The gradients of $\bar{\Delta}_x$ and $\bar{\Delta}_y$ are given by:

$$\begin{aligned}\frac{\partial\bar{\Delta}_x}{\partial\theta_0^f} &= -\sum_{\hat{y}\in\mathcal{Y}} P(\hat{y}|x; F_\gamma)\frac{\delta(x, \hat{y}; F_\alpha, G_\beta, F_\gamma)}{\partial\theta_0^f},\\ \frac{\partial\bar{\Delta}_x}{\partial\theta_0^g} &= -\sum_{\hat{y}\in\mathcal{Y}} P(\hat{y}|x; F_\gamma)\frac{\delta(x, \hat{y}; F_\alpha, G_\beta, F_\gamma)}{\partial\theta_0^g},\\ \frac{\partial\bar{\Delta}_y}{\partial\theta_0^g} &= -\sum_{\hat{x}\in\mathcal{X}} P(\hat{x}|y; G_\gamma)\frac{\delta(y, \hat{x}; G_\beta, F_\alpha, G_\gamma)}{\partial\theta_0^g},\\ \frac{\partial\bar{\Delta}_y}{\partial\theta_0^f} &= -\sum_{\hat{x}\in\mathcal{X}} P(\hat{x}|y; G_\gamma)\frac{\delta(y, \hat{x}; G_\beta, F_\alpha, G_\gamma)}{\partial\theta_0^f},\end{aligned} \tag{4.16}$$

with

$$\delta(x, \hat{y}; F_\alpha, G_\beta, F_\gamma) = \left(\frac{P(\hat{y}|x; F_\alpha)}{P(\hat{y}|x; F_\gamma)}\right) \log P(x|\hat{y}; G_\beta),$$

$$\delta(y, \hat{x}; G_\beta, F_\alpha, G_\gamma) = \left(\frac{P(\hat{x}|y; G_\beta)}{P(\hat{x}|y; G_\gamma)}\right) \log P(y|\hat{x}; F_\alpha),$$

where $\gamma = (0, \frac{1}{N-1}, \cdots, \frac{1}{N-1})$ and F_γ and G_γ represent the combined models from all the pre-trained agents without the target models f_0 and g_0 (see Eqs. (4.13) and (4.14)).

Note that calculating all the four gradient terms (i.e., $\frac{\partial \bar{\Delta}_x}{\partial \theta_0^f}$, $\frac{\partial \bar{\Delta}_x}{\partial \theta_0^g}$, $\frac{\partial \bar{\Delta}_y}{\partial \theta_0^g}$ and $\frac{\partial \bar{\Delta}_y}{\partial \theta_0^f}$) in the above equations needs summing over the space $\mathcal{X}$ and $\mathcal{Y}$, which are exponentially large. Wang et al. [39] approximate the gradients using Monte Carlo method and importance sampling. Take the calculation of $\frac{\partial \bar{\Delta}_x}{\partial \theta_0^f}$ as an example. One samples one $\hat{y} \in \mathcal{Y}$ according to the distribution $P(\hat{y}|x; F_\gamma)$, and uses $\frac{\delta(x,\hat{y}; F_\alpha, G_\beta, F_\gamma)}{\partial \theta_0^f}$ as an approximation to $\frac{\partial \bar{\Delta}_x}{\partial \theta_0^f}$.

To save GPU memory usage and avoid loading multiple models simultaneously, Wang et al. [39] use offline sampling rather than online sampling. Initially $\hat{x}$ and $\hat{y}$ are sampled offline using F_γ and G_γ, their weights $P(\hat{x}|y; F_\gamma)$ and $P(\hat{y}|x; G_\gamma)$ are calculated, and then the target primal and dual models f_0 and g_0 are trained using those offline sampled data. In this way one only loads two models into GPU memory instead of loading $2N$ models simultaneously for online sampling.

Algorithm 2 shows the whole training process of multi-agent dual machine translation. As shown in the algorithm (Line 6), if bilingual data is available, denoted as $\mathcal{B}$, it is easy to leverage both bilingual data and monolingual data in multi-agent dual machine translation.

Algorithm 2 Multi-agent dual learning for machine translation

1: **input:** Monolingual data $\mathcal{M}_X$ and $\mathcal{M}_Y$; learning rate η; f_i and g_i $i \in \{0, 1, \cdots, N-1\}$; mini-batch size K; bilingual data $\mathcal{B}$ if available;
2: Define $\gamma = (0, \frac{1}{N-1}, \cdots, \frac{1}{N-1})$ and $\alpha = \beta = (\frac{1}{N}, \frac{1}{N}, \cdots, \frac{1}{N})$;
3: **repeat**
4: Randomly sample two batches of $B_X \subset \mathcal{M}_X$ and $B_Y \subset \mathcal{M}_Y$, each of size K;
5: Following (4.16) and the related tricks, calculate the gradients of $\tilde{\ell}_{dual}(B_X, B_Y; F_\alpha, G_\beta)$, w.r.t. θ_0^f and θ_0^g; denote them as Grad_{f_0} and Grad_{g_0};
6: If bilingual data is available, sample a batch $B_{XY} \subset \mathcal{B}$ of size K, calculate

$$\text{Grad}_{f_0} \leftarrow \text{Grad}_{f_0} - \frac{1}{K}\nabla_{\theta_0^f} \sum_{(x,y)\in B_{XY}} \log P(y|x; f_0)$$

$$\text{Grad}_{g_0} \leftarrow \text{Grad}_{g_0} - \frac{1}{K}\nabla_{\theta_0^g} \sum_{(x,y)\in B_{XY}} \log P(x|y; g_0)$$

7: Update the parameters: $\theta_0^f \leftarrow \theta_0^f - \eta\, \text{Grad}_{f_0}$, $\theta_0^g \leftarrow \theta_0^g - \eta\, \text{Grad}_{g_0}$;
8: **until** convergence

4.6 Beyond Machine Translation

In addition to machine translation, the dual reconstruction principle has been leveraged to boost many other natural language tasks. We describe several selected ones in this section.

4.6.1 Semantic Parsing

Semantic parsing [41] is the task of mapping a natural language query into a logical form, one type of meaning representation understood by computers, which usually can be executed by an executor to obtain the answers. Deep neural networks based sequence to sequence learning has been the dominant approach for semantic parsing [17]. Similar to neural machine translation, neural semantic parsing also suffers from insufficient labeled training data, since data annotation of semantic parsing is labor-intensive and time-consuming. Especially, the logical form is unfriendly for human annotation. Furthermore, unlike natural language sentences, logical forms are strictly structured, and standard decoding in the sequence to sequence framework is likely to generate invalid or incomplete logical forms.

To address the above two challenges, Cao et al. [5] propose a new framework based on dual learning, in which the primal task is semantic parsing, mapping queries to logic forms, and the dual task is query generation based on logic forms. The two tasks forms a closed loop, as shown in Fig. 4.4. Through dual reconstruction loops, we can leverage unlabeled data (queries or synthesized logical forms) in a more effective way and thus alleviate the problem of lack of annotated data. Cao et al. further propose a validity reward focusing on the surface and semantics of logical forms, which indicates whether the generated logical form is well-formed or not. It involves the prior-knowledge about structures of logical forms predefined in a domain.

Experiments [5] show that the dual learning method achieves state-of-the-art performance at that time with test accuracy 89.1% on the ATIS dataset and gets competitive accuracy on the OVERNIGHT dataset.

Zhu et al. [44] and Su et al. [34] adopt similar ideas for natural language understanding and generation. Cao et al. [6] further extend the above dual learning method and propose a two-stage semantic parsing framework by introducing a pre-training step, which is an unsupervised paraphrase model to convert an unlabeled natural language utterance into the canonical utterance.

4.6.2 Text Style Transfer

Text style transfer, which aims to rephrase a given text in the target style while preserving its semantics, has various application scenarios such as sentiment

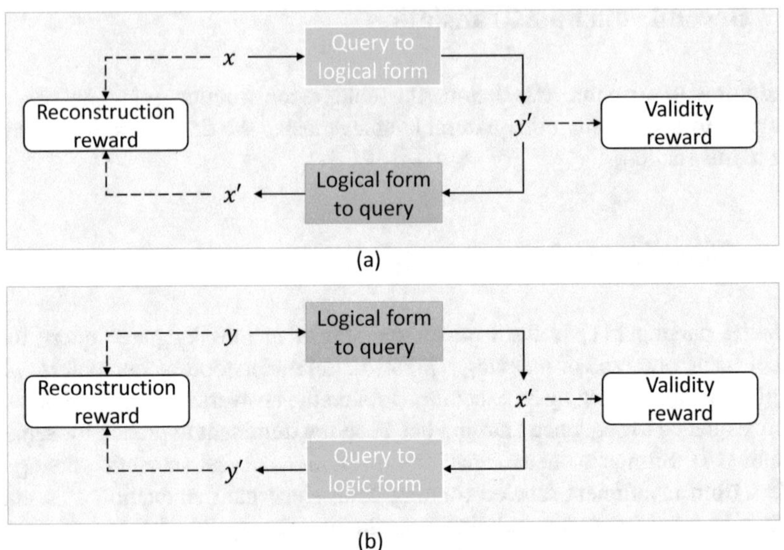

Fig. 4.4 Dual learning for semantic parsing. (**a**) Dual reconstruction loop for query → logical form → query. (**b**) Dual reconstruction loop for logical form → query → logical form

transformation (transferring a positive review to a negative one) and formality modification (revising an informal message into a formal email). As paired data, i.e., paired paragraphs with the same semantics but different styles, is costly to collect, a major approach for the problem is unsupervised learning.

Most methods of unsupervised text style transfer fall in a two-step framework: first separating the semantics from the source style and then fusing the semantics with the target style. The semantics/style separation in the first step is challenging because the semantics and style usually interact in subtle ways in natural language. Luo et al. [23] propose an end-to-end approach to directly transfer the style of the input text to the target style, without explicitly separating semantics and style. They treat the source-to-target style transfer as a primal task and the target-to-source transfer as a dual task, and form a closed loop by transferring a paragraph/sentence with the source style to one with the target style and then back to one with the source style. Two kinds of rewards, the dual reconstruction rewards at the end of the loop and the style classifier rewards inside the loop, are used as feedback signals to enable the learning from unpaired data. Similar to [15], the two mapping models are trained via reinforcement learning, without using paired data.

Automatic evaluations show that the dual learning based method outperforms state-of-the-art systems by a large margin, with more than 8 BLEU points improvement averaged on two benchmark datasets. Human evaluations also validate the effectiveness of the method in terms of both style accuracy and semantics preservation. The code, data, and pre-trained model are available at https://github.com/luofuli/DualLanST.

The principle of dual reconstruction is also used to help fine-grained text sentiment transfer [22].

4.6.3 Conversations

Conversational systems, in which a machine learning model responds to users' inputs/queries, play an important role in many applications, including personal assistant (e.g., Amazon Alexa, Microsoft Cortana, etc.), e-commerce, technical support services, and entertaining chatbots.

It is well-known that personalized conversation models provide better user experiences. A technical challenge for personalized models is that we do not have large scale labeled data to train a personalized model for each individual, since the conversational data collected from an individual user is usually very limited. A straightforward approach is domain adaptation: we first train a general response model using the data collected from public domains and all the users and then fine-tune the general model using user-specific data. Unfortunately, such fine-tuning still suffers from limited data. Yang et al. [40] introduce a dual model, which maps a response back to users' input query, and leverage the dual reconstruction principle to learn from unpaired users' queries and AI's responses.

Today's conversational AIs often generate non-informative responses. To deliver more informative responses, external knowledge sources have been used, which is called Knowledge-Grounded Conversations (KGCs). Meng et al. [24] study the knowledge selection task, a key ingredient in KGC, that aims to select appropriate knowledge to be used in the next response. They design dual knowledge interaction learning, an unsupervised learning scheme based on the dual reconstruction principle to train their proposed neural network, and enable the exploration of extra knowledge besides the knowledge encountered in the training set.

Emotional conversational AIs aim to make open-domain conversations more empathetic and engaging. Shen and Feng [33] propose a framework named Curriculum Dual Learning (CDL) which introduces a dual task of emotional query generation to help the primal task of emotional response generation. CDL utilizes two kinds of rewards that focus on emotion and content respectively for the dual learning process.

Question answering and question generation are related to conversational systems. Actually they can be a module in conversational systems: question answering can respond to the questions asked by users, and question generation can trigger users' interests by actively asking appropriate questions. Zhang and Bansal [42] study the problem of question answering and show that structural duality between the two tasks can help to address the semantic drift in question generation for semi-supervised question answering.

References

1. Artetxe, M., Labaka, G., & Agirre, E. (2017). Learning bilingual word embeddings with (almost) no bilingual data. In *Proceedings of the 55th Annual Meeting of the Association for Computational Linguistics* (pp. 451–462).
2. Artetxe, M., Labaka, G., Agirre, E., & Cho, K. (2018). Unsupervised neural machine translation. In *6th International Conference on Learning Representations*.
3. Bahdanau, D., Cho, K., & Bengio, Y. (2015). Neural machine translation by jointly learning to align and translate. In *3rd International Conference on Learning Representations, ICLR 2015*.
4. Brown, P. F., Cocke, J., Pietra, S. A. D., Pietra, V. J. D., Jelinek, F., Lafferty, J., et al. (1990). A statistical approach to machine translation. *Computational Linguistics, 16*(2), 79–85.
5. Cao, R., Zhu, S., Liu, C., Li, J., & Yu, K. (2019). Semantic parsing with dual learning. In *Proceedings of the 57th Annual Meeting of the Association for Computational Linguistics* (pp. 51–64).
6. Cao, R., Zhu, S., Yang, C., Liu, C., Ma, R., Zhao, Y., et al. (2020). Unsupervised dual paraphrasing for two-stage semantic parsing. Preprint. arXiv:2005.13485.
7. Cho, K., van Merriënboer, B., Gulcehre, C., Bahdanau, D., Bougares, F., Schwenk, H., et al. (2014). Learning phrase representations using RNN encoder–decoder for statistical machine translation. In *Proceedings of the 2014 Conference on Empirical Methods in Natural Language Processing (EMNLP)* (pp. 1724–1734).
8. Conneau, A., Lample, G., Ranzato, M., Denoyer, L., & Jégou, H. (2017). Word translation without parallel data. Preprint. arXiv:1710.04087.
9. Dietterich, T. G. (2002). Ensemble learning. *The Handbook of Brain Theory and Neural Networks, 2* (pp. 110–125). MIT Press.
10. Edunov, S., Ott, M., Auli, M., & Grangier, D. (2018). Understanding back-translation at scale. In *Proceedings of the 2018 Conference on Empirical Methods in Natural Language Processing* (pp. 489–500).
11. Gehring, J., Auli, M., Grangier, D., Yarats, D., & Dauphin, Y. N. (2017). Convolutional sequence to sequence learning. In *Proceedings of the 34th International Conference on Machine Learning* (Vol. 70, pp. 1243–1252). JMLR. org.
12. Gulcehre, C., Firat, O., Xu, K., Cho, K., Barrault, L., Lin, H.-C., et al. (2015). On using monolingual corpora in neural machine translation. Preprint. arXiv:1503.03535.
13. Gulcehre, C., Firat, O., Xu, K., Cho, K., & Bengio, Y. (2017). On integrating a language model into neural machine translation. *Computer Speech & Language, 45*, 137–148.
14. Hassan Awadalla, H., Aue, A., Chen, C., Chowdhary, V., Clark, J., Federmann, C., et al. (March 2018). Achieving human parity on automatic chinese to English news translation. arXiv:1803.05567.
15. He, D., Xia, Y., Qin, T., Wang, L., Yu, N., Liu, T.-Y., et al. (2016). Dual learning for machine translation. In *Advances in Neural Information Processing Systems* (pp. 820–828).
16. Hinton, G., Vinyals, O., & Dean, J. (2015). Distilling the knowledge in a neural network. Preprint. arXiv:1503.02531.
17. Jia, R., & Liang, P. (2016). Data recombination for neural semantic parsing. In *Proceedings of the 54th Annual Meeting of the Association for Computational Linguistics (Volume 1: Long Papers)* (pp. 12–22).
18. Kim, Y., & Rush, A. M. (2016). Sequence-level knowledge distillation. In *Proceedings of the 2016 Conference on Empirical Methods in Natural Language Processing* (pp. 1317–1327).
19. Koehn, P. (2009). *Statistical machine translation*. New York: Cambridge University Press.
20. Lample, G., Conneau, A., Denoyer, L., & Ranzato, M. (2018). Unsupervised machine translation using monolingual corpora only. In *6th International Conference on Learning Representations, ICLR 2018*.
21. Lample, G., Ott, M., Conneau, A., Denoyer, L., & Ranzato, M. (2018). Phrase-based & neural unsupervised machine translation. In *Proceedings of the 2018 Conference on Empirical*

Methods in Natural Language Processing, Brussels, Belgium, October 31–November 4, 2018 (pp. 5039–5049).
22. Luo, F., Li, P., Yang, P., Zhou, J., Tan, Y., Chang, B., et al. (2019). Towards fine-grained text sentiment transfer. In *Proceedings of the 57th Annual Meeting of the Association for Computational Linguistics* (pp. 2013–2022).
23. Luo, F., Li, P., Zhou, J., Yang, P., Chang, B., Sun, X., et al. (2019). A dual reinforcement learning framework for unsupervised text style transfer. In *Proceedings of the 28th International Joint Conference on Artificial Intelligence* (pp. 5116–5122). AAAI Press.
24. Meng, C., Ren, P., Chen, Z., Sun, W., Ren, Z., Tu, Z., et al. (2020). Dukenet: A dual knowledge interaction network for knowledge-grounded conversation. In *Proceedings of the 43rd International ACM SIGIR Conference on Research and Development in Information Retrieval* (pp. 1151–1160).
25. Mikolov, T., Karafiát, M., Burget, L., Černocký, J., & Khudanpur, S. (2010). Recurrent neural network based language model. In *Eleventh Annual Conference of the International Speech Communication Association.*
26. Mikolov, T., Sutskever, I., Chen, K., Corrado, G. S., & Dean, J. (2013). Distributed representations of words and phrases and their compositionality. In *Advances in Neural Information Processing Systems* (pp. 3111–3119).
27. Nirenburg, S. (1989). Knowledge-based machine translation. *Machine Translation, 4*(1), 5–24.
28. Nirenburg, S., Carbonell, J., Tomita, M., & Goodman, K. (1994). *Machine translation: A knowledge-based approach.* San Mateo, CA: Morgan Kaufmann Publishers Inc.
29. Ranzato, M., Chopra, S., Auli, M., & Zaremba, W. (2015) Sequence level training with recurrent neural networks. Preprint. arXiv:1511.06732.
30. Rush, A. M., Chopra, S., & Weston, J. (2015). A neural attention model for abstractive sentence summarization. In *Proceedings of the 2015 Conference on Empirical Methods in Natural Language Processing* (pp. 379–389).
31. Sennrich, R., Haddow, B., & Birch, A. (2016). Improving neural machine translation models with monolingual data. In *Proceedings of the 54th Annual Meeting of the Association for Computational Linguistics (Volume 1: Long Papers)* (pp. 86–96).
32. Sestorain, L., Ciaramita, M., Buck, C., & Hofmann, T. (2018). Zero-shot dual machine translation. Preprint. arXiv:1805.10338.
33. Shen, L., & Feng, Y. (2020). CDL: Curriculum dual learning for emotion-controllable response generation. Preprint. arXiv:2005.00329.
34. Su, S.-Y., Huang, C.-W., & Chen, Y.-N. (2020). Towards unsupervised language understanding and generation by joint dual learning. In *ACL 2020: 58th Annual Meeting of the Association for Computational Linguistics* (pp. 671–680).
35. Sundermeyer, M., Schlüter, R., & Ney, H. (2012). LSTM neural networks for language modeling. In *Thirteenth Annual Conference of the International Speech Communication Association.*
36. Sutskever, I., Vinyals, O., & Le, Q. V. (2014). Sequence to sequence learning with neural networks. In *Advances in Neural Information Processing Systems* (pp. 3104–3112).
37. Sutton, R. S., McAllester, D. A., Singh, S. P., & Mansour, Y. (2000). Policy gradient methods for reinforcement learning with function approximation. In *Advances in Neural Information Processing Systems,* (pp. 1057–1063).
38. Vaswani, A., Shazeer, N., Parmar, N., Uszkoreit, J., Jones, L., Gomez, A. N., et al. (2017). Attention is all you need. In *Advances in Neural Information Processing Systems* (pp. 5998–6008).
39. Wang, Y., Xia, Y, He, T., Tian, F., Qin, T., Xiang Zhai, C., et al. (2019). Multi-agent dual learning. In *7th International Conference on Learning Representations, ICLR 2019.*
40. Yang, M., Zhao, Z., Zhao, W., Chen, X., Zhu, J., Zhou, L., et al. (2017). Personalized response generation via domain adaptation. In *Proceedings of the 40th International ACM SIGIR Conference on Research and Development in Information Retrieval* (pp. 1021–1024).

41. Zelle, J. M., & Mooney, R. J. (1996). Learning to parse database queries using inductive logic programming. In *Proceedings of the National Conference on Artificial Intelligence* (pp. 1050–1055).
42. Zhang, S., & Bansal, M. (2019). Addressing semantic drift in question generation for semi-supervised question answering. In *Proceedings of the 2019 Conference on Empirical Methods in Natural Language Processing and the 9th International Joint Conference on Natural Language Processing (EMNLP-IJCNLP)* (pp. 2495–2509).
43. Zhou, Z.-H. (2012). *Ensemble methods: Foundations and algorithms*. New York: CRC Press.
44. Zhu, S., Cao, R., & Yu, K. (2020). Dual learning for semi-supervised natural language understanding. *IEEE Transactions on Audio, Speech, and Language Processing, 28*, 1936–1947.

Chapter 5
Dual Learning for Image Translation and Beyond

5.1 Introduction

Conditional image generation, which is an important problem in image processing, computer vision and graphics, has made rapid progress and achieved great success in recent years powered by deep learning techniques. Especially after generative adversarial networks (GANs) have been proposed [5], many variants of GANs have been designed for image generation conditioned on different kinds of inputs, including class labels, attributes, texts, and images.

Image to image translation (or image translation for short), which "translates" an image to a corresponding output image, is a kind of image generation conditioned on images and covers many tasks, such as super resolution, texture synthesis, image inpainting, gray image colorization, image style/domain transfer, etc. While many algorithms/models have been designed for specific applications of image translation, a recent trend is to design general-purpose algorithms, which can be categorized into two major kinds of approaches: supervised image translation and unsupervised image translation.

Supervised image translation takes paired images as inputs to learn a translation model. An image pair contains an input image and its corresponding output image, as shown in Fig. 5.1. Algorithms within this supervised framework (e.g., conditional GAN [8]) usually perform well in terms of translation quality, but suffer from high labeling cost and therefore are costly to scale to many domains.

Unsupervised algorithms [2, 9, 16, 33, 39] for image translation attract more and more research attentions recently, in which dual learning plays a key role. In this chapter, we focus on dual learning for unsupervised image translation and introduce several representative algorithms for unconditional image translation (Sect. 5.3) and conditional image translation (Sect. 5.4).

T. Qin, *Dual Learning*, https://doi.org/10.1007/978-981-15-8884-6_5

Fig. 5.1 Paired images for different image translation tasks. Figure reproduced with permission from Isola et al. [8]

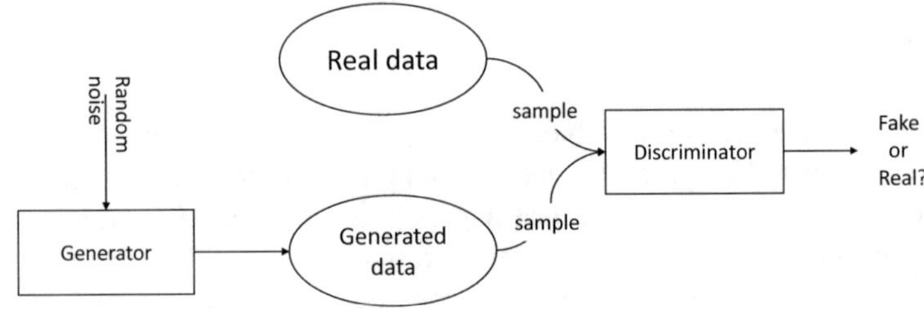

Fig. 5.2 Basic idea of GANs

5.1.1 Generative Adversarial Networks

Generative adversarial networks (GANs) [5] are a machine learning framework for generative model training. The key of GANs is the introduction of a discriminative model, which estimates the quality of samples generated from a generative model by classifying whether samples are real or fake.

As shown in Fig. 5.2, a GAN has two major components: a generative model G that generates samples from random noise, and a discriminative model D that discriminates generated ones from real ones. The training process of GANs corresponds to a two-player minmax game: the goal of the discriminator D is to minimize the classification error, while the generator G is to maximize the probability of the discriminator D making a mistake. If we do not add any constraints on the generator and discriminator, there exists a unique solution for the game, in which G perfectly captures the training data distribution and D outputs $\frac{1}{2}$ everywhere. That is, if we do not consider the optimization difficulty of the training process and assume that the expressive power of the models are unlimited, eventually the generator G will recover the true distribution of the training data and the discriminator cannot differentiate the generated samples from real samples.

GANs have many variants and extensions [1, 6, 8, 22, 24, 30, 39], discussing about how to improve the training process of GANs and apply to other problems. Readers can check recent survey papers [4, 29] for better understanding.

5.2 Basic Idea of Unsupervised Image Translation

While there are many algorithms for unsupervised image to image translation [9, 16, 17, 33, 39], they share the basic idea: combining dual reconstruction and GANs. We first define some notations and then introduce the basic idea.

Let $\mathcal{D}_X$ denote the collection of images from domain X, and $\mathcal{D}_Y$ the collection of images from domain Y. There no correspondence between the images in $\mathcal{D}_X$ and $\mathcal{D}_Y$. Let $\theta_{XY}()$ denote the forward/primal generator that translates an image from domain X to Y, and $\theta_{YX}()$ the backward/dual generator that from Y to X. When the context is clear and there is no confusion, we reuse θ_{XY} and θ_{YX} to denote the parameters of the two generators.

As unsupervised image translation aims to learn a translation model using the unpaired images from $\mathcal{D}_X$ and $\mathcal{D}_Y$, there is no direct feedback signal in the data for model training. Therefore, almost all the algorithms for unsupervised image translation jointly train the primal and dual translators, and take the reconstruction error as feedback signal following the dual reconstruction principle (see Sect. 4.2). That is, the training process is to minimize the dual reconstruction error

$$\min_{\theta_{XY},\theta_{YX}} \Delta(x, \theta_{YX}(\theta_{XY}(x)), \forall x \in \mathcal{D}_X, \tag{5.1}$$

and

$$\min_{\theta_{XY},\theta_{YX}} \Delta(y, \theta_{XY}(\theta_{YX}(y)), \forall y \in \mathcal{D}_Y. \tag{5.2}$$

Unfortunately, only the dual reconstruction principle is not enough to guarantee that the learnt translation models are meaningful. For example, a trivial solution of the above minimization problems is the copy operator: $\theta_{XY}(x) = x$ and $\theta_{YX}(y) = y$, which leads to zero dual reconstruction error. However, such two copy models do not achieve the goal of translating an image from a source domain to a target domain; the output of the models is still in the source domain.

To avoid this kind of trivial solutions and ensure that the two translation models really translate an image from one domain to the other domain, inspired by the idea of adversary training in GANs [5], two discriminators are introduced into image translation. Let θ_X denote the discriminator for domain X, which takes an image as input and outputs the probability of the image from domain X, and θ_Y the discriminator for domain B. Similarly, we reuse θ_X and θ_Y to denote the parameters of the two discriminators.

θ_X is trained to maximize the probability $\theta_X(x)$ of a natural image x from domain X and minimize the probability $\theta_X(\theta_{YX}(y))$ of an image $\theta_{YX}(y)$ generated by the dual translation model θ_{YX} from image y. Similarly, θ_Y is trained to maximize the probability $\theta_Y(y)$ of a natural image y from domain Y and minimize the probability $\theta_Y(\theta_{XY}(x))$ of an image $\theta_{XY}(x)$ generated by the primal translation model θ_{XY} from image x.

In contrast, the training of the two translation models needs to fool the two discriminators as follows.

$$\max_{\theta_{XY},\theta_{YX}} \theta_X(\theta_{YX}(y)), \forall y \in \mathcal{D}_Y \tag{5.3}$$

$$\max_{\theta_{XY},\theta_{YX}} \theta_Y(\theta_{XY}(x)), \forall x \in \mathcal{D}_X \tag{5.4}$$

While the introduction of adversarial training (or GANs) can help dual reconstruction to avoid trivial solution, dual reconstruction also helps GANs to remedy the mode-collapsing problem [21] to some extent. Mode collapsing is the failure of GANs that the learnt generator collapses to producing only a single image or a small family of very similar images for diverse inputs. As dual reconstruction aims to minimize reconstruction error, a mode-collapsing generator will be discouraged: if a primal generator collapses to a single image, its dual generator cannot reconstruct the diverse inputs of the primal generator, and therefore the dual reconstruction error of such a pair of generators will be large.

5.3 Image to Image Translation

In this section, we focus on image to image translation in two domains, and introduce three representative works: DualGAN [33], CycleGAN [39] and DiscoGAN [9].

5.3.1 DualGAN

In this sub section, we introduce the DualGAN model, including its loss functions, network architectures and training procedure.

Figure 5.3 shows the data flow of DualGAN. As can be seen, DualGAN exactly follows the dual reconstruction principle and the GAN-style principle.

The two translation models are trained to minimize the dual reconstruction error between a natural image x or y and its reconstructed copy $\theta_{YX}(\theta_{XY}(x, z), z')$ or

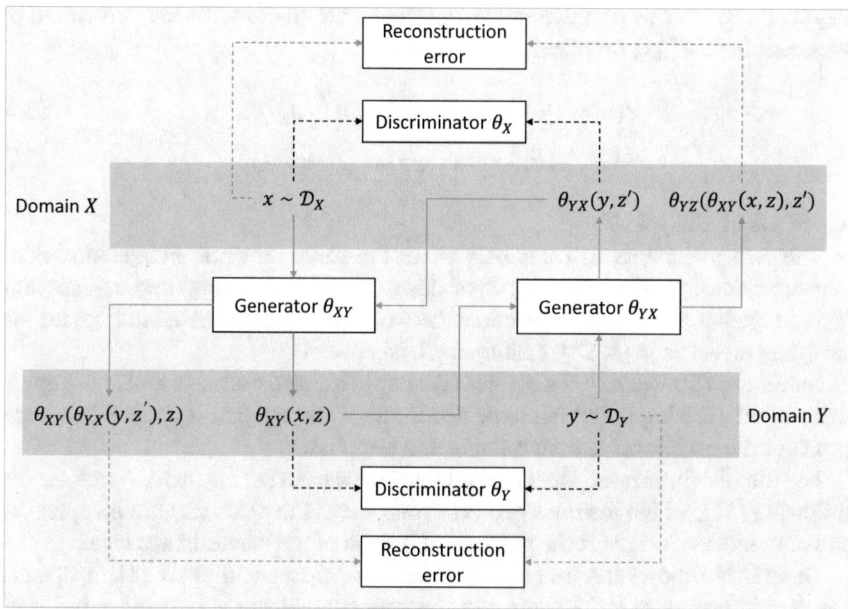

Fig. 5.3 DualGAN for image-to-image translation. Yellow solid arrows make up the dual reconstruction loop starting from domain X, blue solid arrows make up the dual reconstruction loop starting from domain Y, yellow and blue dashed arrows make up the dual reconstruction error, and black dashed arrows make up the discriminator loss

$\theta_{XY}(\theta_{YX}(y, z'), z)$ and maximize the error probability of the discriminators.[1] In particular, DualGAN adopts L_1 distance instead of L_2 distance to measure the dual construction error, considering that L_2 distance often leads to blurriness [10]. Then the training loss of the two translation models is

$$\begin{aligned} \ell(x, y; \theta_{YX}, \theta_{XY}) &= \lambda_X \|x - \theta_{YX}(\theta_{XY}(x, z), z')\| \\ &+\lambda_Y \|y - \theta_{XY}(\theta_{YX}(y, z'), z)\| \\ &-\theta_Y(\theta_{XY}(x, z)) - \theta_X(\theta_{YX}(y, z')). \end{aligned} \tag{5.5}$$

Here x and y are two unpaired images from domain X and Y separately, and λ_X and λ_Y are two hyper-parameters. Yi et al. [33] suggest setting λ_X and λ_Y to a value within [100.0, 1000.0]. They find that if $\mathcal{X}$ contains natural images and $\mathcal{Y}$ does not (e.g., aerial photo-maps), it is preferred to use smaller λ_X than λ_Y.

The discriminator θ_X is trained with $\theta_{YX}(y, z'), \forall y \in \mathcal{Y}$ as negative examples and $x \in \mathcal{X}$ as positive examples, while θ_Y with $y \in \mathcal{Y}$ as positive examples and

[1]Note that z and z' are random noise introduced into the translation process.

$\theta_{XY}(x, z), \forall x \in \mathcal{X}$ as negative examples. DualGAN employs the loss advocated by Wasserstein GAN [1] for discriminator training:

$$\ell(x, y; \theta_X) = \theta_X(\theta_{YX}(y, z')) - \theta_X(x), \tag{5.6}$$

$$\ell(x, y; \theta_Y) = \theta_Y(\theta_{XY}(x, z)) - \theta_Y(y), \tag{5.7}$$

where $x \in \mathcal{X}$ and $y \in \mathcal{Y}$.

The two generators in DualGAN share the same network architecture. Each generator consists of equal number of down-sampling (pooling) and up-sampling layers together with skip connections between mirrored down-sampling and up-sampling layers as in [8, 23], making it a U-shaped net.

Similar to [8], DualGAN does not take explicit noise vectors z and z' as inputs. Instead, they are implicitly implemented through dropout that is applied to several layers of the generators at both training and test phases.

For the discriminators, DuaGAN adopts the Markovian PatchGAN architecture following [11], which assumes pixels beyond a specific patch size are independent and only models images at the patch level instead of the whole-image level.

DualGAN follows the training procedure of Wasserstein GAN [1]. It updates the discriminators every d steps and the generators every step using mini-batch stochastic gradient descent with the RMSProp optimizer. d is set to 2–4 and batch size m is set to 1–4 in [33]. The clipping parameter c is set in $[0.01, 0.1]$, depending on tasks. The details of the training procedure can be found at Algorithm 1.

Algorithm 1 DualGAN training procedure

Require: Two image collections $\mathcal{D}_X$ and $\mathcal{D}_Y$, clipping parameter c, batch size m, and d

Randomly initialize $\theta_{YX}, \theta_{XY}, \theta_X$, and θ_Y
repeat
  for $t = 1, \ldots, d$ **do**
    Sample images $\{x^{(k)}\}_{k=1}^m \subseteq \mathcal{D}_X, \{y^{(k)}\}_{k=1}^m \subseteq \mathcal{D}_Y$
    Update θ_X to minimize $\frac{1}{m}\sum_{k=1}^m \ell(x^{(k)}, y^{(k)}; \theta_X)$
    Update θ_Y to minimize $\frac{1}{m}\sum_{k=1}^m \ell(x^{(k)}, y^{(k)}; \theta_Y)$
    clip$(\theta_X, -c, c)$, *clip*$(\theta_Y, -c, c)$
  end for
  Sample images $\{x^{(k)}\}_{k=1}^m \subseteq \mathcal{D}_X, \{y^{(k)}\}_{k=1}^m \subseteq \mathcal{D}_Y$
  Update θ_{YX} and θ_{XY} to minimize $\frac{1}{m}\sum_{k=1}^m \ell(x^{(k)}, y^{(k)}; \theta_{YX}, \theta_{XY})$
until convergence

5.3.2 CycleGAN

CycleGAN [39] also combines the principle of dual reconstruction and adversarial training, and is one of the most popular algorithms for unpaired image to image translation.

Similar to DualGAN, CycleGAN also trains two generators θ_{YX} and θ_{XY} together, to leverage the dual structure of two translation tasks and obtain feedback signal for model training based on the dual reconstruction principle.[2] Different from DualGAN, in which the two generators take an image and a noise vector as inputs (see Eq. (5.5)), the two generators $\theta_{XY}()$ and $\theta_{YX}()$ take only an image as input (see Eq. (5.8)). CycleGAN uses the same dual reconstruction objective as DualGAN:

$$\begin{aligned} \ell_{dual}(\theta_{XY}, \theta_{YX}) = & E_{x\sim\mathcal{X}}\|x - \theta_{YX}(\theta_{XY}(x))\| \\ & + E_{y\sim\mathcal{Y}}\|y - \theta_{XY}(\theta_{YX}(y))\|, \end{aligned} \tag{5.8}$$

where $\mathcal{X}$ and $\mathcal{Y}$ denote two image domains. Zhu et al. [39] also tried replacing the L1 norm in this objective with an adversarial objective between x and $\theta_{YX}(\theta_{XY}(x))$, and between y and $\theta_{XY}(\theta_{YX}(y))$, but no improvement was observed.

CycleGAN also introduces two discriminators θ_X and θ_Y, to ensure that a generator indeed translates an image from one domain to the other domain and the generated image is indistinguishable from real images from the other domain. While in general the adversarial objective is defined as

$$\begin{aligned} \ell_{adv}(\theta_{XY}, \theta_{YX}, \theta_X, \theta_Y) = & E_{x\sim\mathcal{X}}[\log(1 - \theta_Y(\theta_{XY}(x)))] \\ & + E_{y\sim\mathcal{Y}}[\log(1 - \theta_X(\theta_{YX}(y)))] \\ & + E_{y\sim\mathcal{Y}}[\log\theta_Y(y)] + E_{x\sim\mathcal{X}}[\log\theta_X(x)], \end{aligned} \tag{5.9}$$

CycleGAN replaces the negative log likelihood objective by a least-square loss [20] and obtains the following objective to stabilize the training procedure.

$$\begin{aligned} \ell_{adv}(\theta_{XY}, \theta_{YX}, \theta_X, \theta_Y) = & E_{x\sim\mathcal{X}}[(1 - \theta_Y(\theta_{XY}(x)))^2] \\ & + E_{y\sim\mathcal{Y}}[(1 - \theta_X(\theta_{YX}(y)))^2] \\ & + E_{y\sim\mathcal{Y}}[\theta_Y(y)^2] + E_{x\sim\mathcal{X}}[\theta_X(x)^2] \end{aligned} \tag{5.10}$$

[2] Zhu et al. [39] use a different name, "cycle consistence", which is previously used in computer vision community [38] and shares exactly the same principle as dual reconstruction. This is why their approach is called "CycleGAN".

Then the overall objective of CycleGAN becomes

$$\begin{aligned}\ell(\theta_{XY}, \theta_{YX}, \theta_X, \theta_Y) = &\ \ell_{dual}(\theta_{XY}, \theta_{YX}) \\ &+ \lambda \ell_{adv}(\theta_{XY}, \theta_{YX}, \theta_X, \theta_Y),\end{aligned} \tag{5.11}$$

where $\ell_{dual}(\theta_{XY}, \theta_{YX})$ is the dual reconstruction objective defined in Eq. (5.8), $\ell_{adv}(\theta_{XY}, \theta_{YX}, \theta_X, \theta_Y)$ is the adversarial objective defined in Eq. (5.10), and λ is a hyper parameter that trades off the two objectives. Then the two generators are obtained through solving the minmax problem as below:

$$\min_{\theta_{XY}, \theta_{YX}} \max_{\theta_X, \theta_Y} \ell(\theta_{XY}, \theta_{YX}, \theta_X, \theta_Y). \tag{5.12}$$

The ablations studies in [39] empirically show that

- Both objectives are critical to translate and generate high-quality images; using only one objective leads to low-quality translations.
- For the dual reconstruction objective, using only unidirectional reconstruction (e.g., only $\|x - \theta_{YX}(\theta_{XY}(x))\|$ or only $\|y - \theta_{XY}(\theta_{YX}(y))\|$) is not sufficient to regularize the training procedure and leads to low-quality translations.

CycleGAN is very popular. Interested readers can find the open-source code, pre-trained models, and amazing results at https://github.com/junyanz/CycleGAN. Figure 5.4 shows the results od CycleGAN, including translations between Monet paintings and landscape photos, translations between zebras and horses, translations summer photos and winter photos, and translations from a natural photo to paintings with different styles.

CycleGAN only uses one forward translator and one backward translator for image translation. Wang et al. [28] apply the multi-agent dual learning framework to image translation and, and introduce multiple forward and backward translators to enhance CycleGAN. The same idea can also be applied to DualGAN and DiscoGAN.

5.3.3 DiscoGAN

DiscoGAN [9] is proposed to discover relations between different domains. It follows the basic idea described in Sect. 5.2 and shares the same training objectives as DualGAN. To avoid redundancy, we omit its technical details in this section. Instead, we would like to show an interesting point of DiscoGAN.

DualGAN and CycleGAN focus on translations between two visually similar image domains. For example, DualGAN conducts experiments on translation between day and night images, label and facade images, photos and sketches, Chinese and oil paintings, etc. CycleGAN conducts experiments on translation between photos and paintings, summer and winter images, zebras and horses, etc.

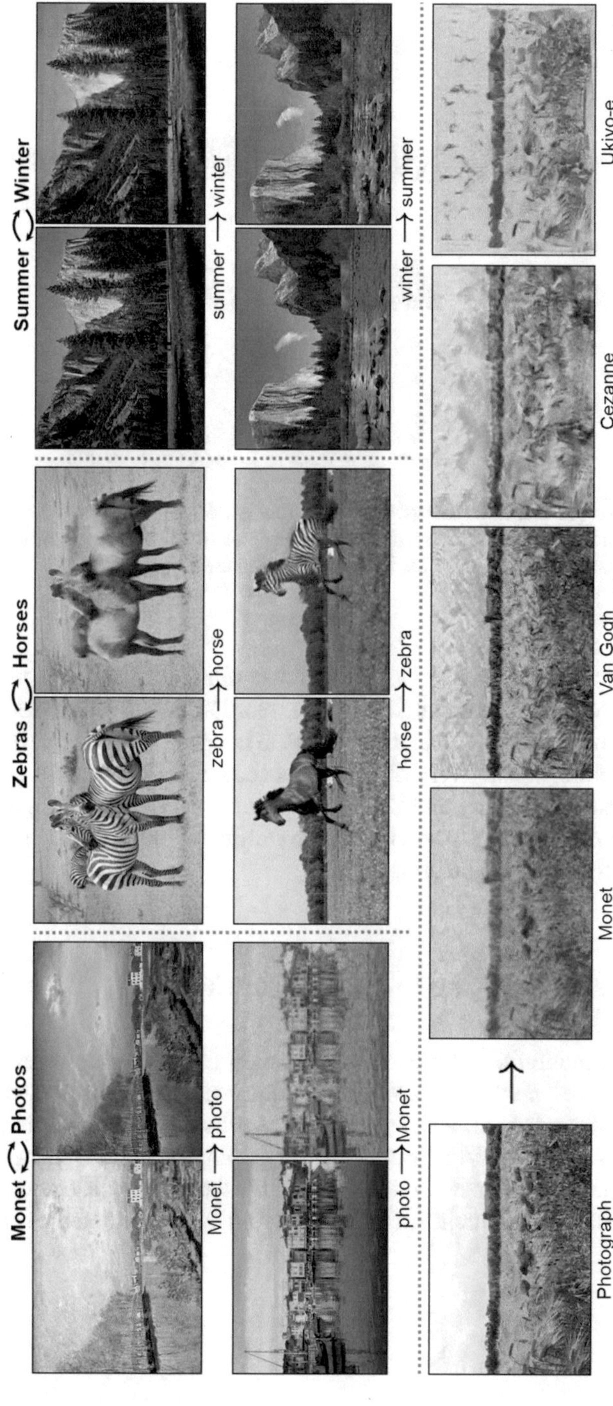

Fig. 5.4 Several translation results of CycleGAN. Figure reproduced with permission from Zhu et al. [39]

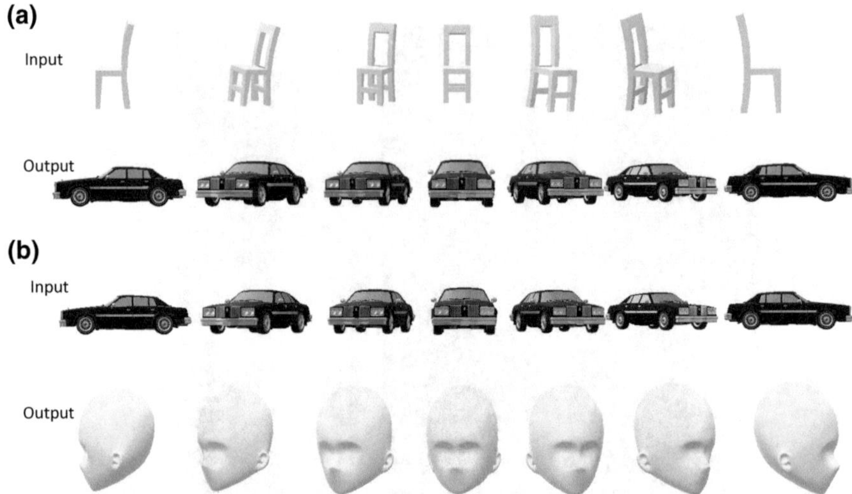

Fig. 5.5 Illustration of DiscoGAN results. (**a**) Translation between chairs and cars. The model was trained on chair and car images. (**b**) Translation between cars and faces. The model was trained on car and face images. Note that here the figures are human created for better illustration but not algorithmic results in [9]

The source and target domains contain visually similar or related objects, and the major changes happen in color, texture, style, etc. Different from DualGAN and CycleGAN, DiscoGAN shows that it can translate images from a domain to another domain that may be visually very different. As shown in Fig. 5.5, DiscoGAN can do translation between chairs and cars, and between cars and faces, surprisingly discover relations of images from visually very different objects, and successfully pair images with similar orientation.

5.4 Fine-Grained Image to Image Translation

An implicit assumption of image-to-image translation is that an image contains two kinds of features: *domain-independent features*, which are preserved during the translation (i.e., the edges of face, eyes, nose and mouse while translating a man's photo to a woman's photo), and *domain-specific features*, which are changed during the translation (i.e., the color and style of the hair for face image translation). Image-to-image translation aims at transferring images from the source domain to the target

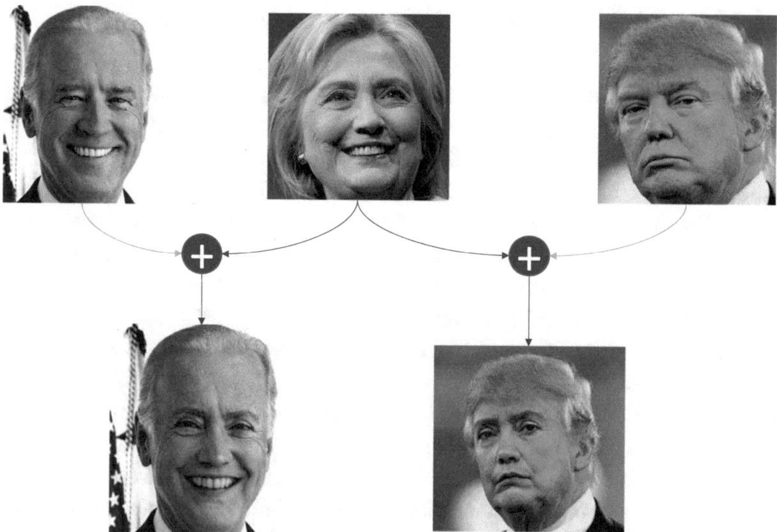

Fig. 5.6 The purple arrow represents the translation flow and the green arrow represents the conditional information flow

domain by preserving domain-independent features while replacing domain-specific features.[3]

While the algorithms introduced in the previous section can successfully translate an image from a source domain to a target domain, a limitation is that they cannot control or manipulate the style in fine granularity of the generated image in the target domain. Consider task of translating a man's photo to a woman's as studied in [9]. Can we translate Hillary's photo to a man' photo with the hair style and color of Trump? DiscoGAN [9] can indeed generate a woman's photo given a man's photo as input, but cannot control the hair style or color of the generated image, neither DualGAN nor CycleGAN.

To enable this kind of fine-granularity control, Lin et al. [16] define and study the task conditional image-to-image translation, which can control domain-specific features in the target domain, specified by another image from the target domain. An example of conditional image-to-image translation is shown in Fig. 5.6, in which we want to convert Hillary's photo to a man's photo. As shown in the figure, with an additional man's photo as input, we can control the translated image (e.g., the hair color and style).

[3]Note that the two kinds of features are relative concepts, and domain-specific features in one task might be domain-independent features in another task.

5.4.1 The Problem of Fine-Grained Image Translation

Following the implicit assumption, an image $x_A \in \mathcal{D}_A$ can be decomposed into two kinds of features: $x_A = x_A^i \oplus x_A^s$, where x_A^i is the domain-independent features, x_A^s is the domain-specific features, and $\oplus$ is the operator that can merge the two kinds of features into a complete image.[4] Similarly, for an image $x_B \in \mathcal{D}_B$, we have $x_B = x_B^i \oplus x_B^s$. Take the images in Fig. 5.6 as examples: (1) If the two domains are man's and woman's photos, the domain-independent features are individual facial organs like eyes and mouths and the domain-specific features are beard and hair style; (2) if the two domains are real bags and the edges of bags, the domain-independent features are exactly the edges of bags themselves, and the domain-specific are the colors and textures.

With above notations, Lin et al. [16] define the task of conditional image-to-image translation: Given an main input image $x_A \in \mathcal{D}_A$ and a conditional input image $x_B \in \mathcal{D}_B$, generate an image x_{AB} in domain $\mathcal{D}_B$ that keeps the domain-independent features of x_A and combines the domain-specific features carried in x_B. Mathematically, the conditional translation task is

$$x_{AB} = G_{A\to B}(x_A, x_B) = x_A^i \oplus x_B^s. \tag{5.13}$$

Note that here the generator/translator $G_{A\to B}()$ takes two image as inputs instead of the single image input as introduced in Sect. 5.3. Similarly, we have the dual/backward conditional translation

$$x_{BA} = G_{B\to A}(x_B, x_A) = x_B^i \oplus x_A^s. \tag{5.14}$$

5.4.2 Conditional DualGAN

Lin et al. [16] design a model for conditional image to image translation based on DualGAN, which is named as cd-GAN. Figure 5.7 shows the overall architecture and the training objective of the proposed model, in which the left part is an encoder-decoder based framework for image translation and the right part shows additional components introduced for model training.

As shown in the figure, cd-GAN consists of two encoders e_A and e_B and two decoders g_A and g_B for the two domains respectively. The encoders serve as feature extractors, which take an image as input and output domain-independent features and domain-specific features. In particular, given two images x_A and x_B, we have

$$(x_A^i, x_A^s) = e_A(x_A), \quad (x_B^i, x_B^s) = e_B(x_B).$$

[4]Here $\oplus$ is just a virtual symbol defined to better understand the concepts of two kinds of features. A generator will be trained from data to generate an image based on two kinds of features.

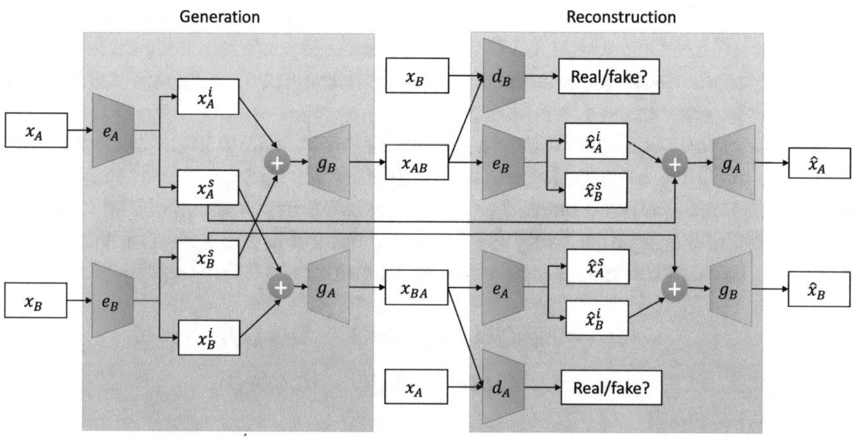

Fig. 5.7 Architecture and training objective of Conditional DuaGAN in [16]

The decoders serve as generators, which take as inputs the domain-independent features from the image in the source domain and the domain-specific features from the image in the target domain and generate an image in the target domain. That is,

$$x_{AB} = g_B(x_A^i, x_B^s), \quad x_{BA} = g_A(x_B^i, x_A^s).$$

The training of cd-GAN is also based on the dual reconstruction principle and the adversarial training principle.

To reconstruct the two images $\hat{x}_A$ and $\hat{x}_B$, as shown in Fig. 5.7, cd-GAN first extracts the two kinds of features of the generated images:

$$(\hat{x}_A^i, \hat{x}_B^s) = e_B(x_{AB}), \quad (\hat{x}_B^i, \hat{x}_A^s) = e_A(x_{BA}),$$

and then reconstructs images:

$$\hat{x}_A = g_A(\hat{x}_A^i, x_A^s), \quad \hat{x}_B = g_B(\hat{x}_B^i, x_B^s).$$

The dual reconstruction quality is evaluated from three aspects: the image-level reconstruction error $\ell_{\text{dual}}^{\text{im}}$, the reconstruction error $\ell_{\text{dual}}^{\text{di}}$ of the domain-independent features, and the reconstruction error $\ell_{\text{dual}}^{\text{ds}}$ of the domain-specific features:

$$\ell_{\text{dual}}^{\text{im}}(x_A, x_B) = \|x_A - \hat{x}_A\|^2 + \|x_B - \hat{x}_B\|^2,$$

$$\ell_{\text{dual}}^{\text{di}}(x_A, x_B) = \|x_A^i - \hat{x}_A^i\|^2 + \|x_B^i - \hat{x}_B^i\|^2,$$

$$\ell_{\text{dual}}^{\text{ds}}(x_A, x_B) = \|x_A^s - \hat{x}_A^s\|^2 + \|x_B^s - \hat{x}_B^s\|^2.$$

Rather than only considering the image-level reconstruction error as in DualGAN/CycleGAN/DiscoGAN, cd-GAN takes more aspects into account and therefore is expected to achieve better accuracy. Another minor difference is that L2 error instead of L1 error is used in cd-GAN.

To ensure the generated images x_{AB} and x_{BA} look natural in the corresponding domains, cd-GAN introduces two discriminators d_A and d_B to differentiate the real images and synthetic ones. d_A (or d_B) takes an image as input and outputs a probability indicating how likely the input is a natural image from domain $\mathcal{D}_A$ (or $\mathcal{D}_B$). The two discriminators are trained to maximize the following objective:

$$\begin{aligned}\ell_{\text{adv}} =& \log(d_A(x_A)) + \log(1 - d_A(x_{BA})) \\ &+ \log(d_B(x_B)) + \log(1 - d_B(x_{AB})),\end{aligned}$$

while the two translators (the encoders e_A and e_B and the decoders g_A and g_B) are trained to minimize the above adversarial objective and the three dual reconstruction objectives simultaneously.

5.4.3 *Discussions*

We make some discussions about cd-GAN in this sub section.

First, we explain why x_A^i and x_B^i are domain independent and x_A^s and x_B^s are domain specific. Consider the path of $x_A \to e_A \to x_A^i \to g_B \to x_{AB}$ in Fig. 5.7. Suppose after training the two translators are of high quality and the generated image x_{AB} is indeed indistinguishable from real images in domain B.

- Note that x_{AB} is generated by combining x_A^i and x_B^s and x_A^i is extracted from an image in domain A. That is, x_A^i is inherited from an image from domain A and preserved in an image in domain B. Thus, it should be domain independent; otherwise, x_A^i carries information about domain A and x_{AB} will look like a natural image in domain A.
- Since x_A^i comes from domain A and is not relevant to domain B, x_B^s must carry information about domain B; otherwise, x_{AB} will not look like a natural image in domain B. Thus, x_B^s is domain specific.
- Similarly, we can see that x_A^s is domain-specific and x_B^i is domain-independent.

Second, DualGAN, CycleGAN, and DiscoGAN introduced in the previous section can be treated as simplified versions of cd-GAN, by removing the domain-specific features. For example, in CycleGAN, given an $x_A \in \mathcal{D}_A$, any $x_{AB} \in \mathcal{D}_B$ is a legal translation, no matter what $x_B \in \mathcal{D}_B$ is. The cd-GAN model requires that the generated images should match the inputs from two domains, which is more difficult.

Third, cd-GAN works for both symmetric translations and asymmetric translations. In symmetric translations, both directions of translations need conditional inputs (see Fig. 5.6a). In asymmetric translations, only one direction of translation needs a conditional image as input (see Fig. 5.6b). That is, the translation from bag to edge does not need another edge image as input; even if an additional edge image is provided as the conditional input, it does not help the translation or change the translation result.

For asymmetric translations, one needs to slightly modify objectives for cd-GAN training. Suppose the translation direction of $G_{B\to A}$ does not need a conditional input. Then cd-GAN does not need to reconstruct the domain-specific features x_A^s. Accordingly, the reconstruction error of domain-specific features becomes

$$\ell_{\text{dual}}^{\text{ds}}(x_A, x_B) = \|x_B^s - \hat{x}_B^s\|^2,$$

and other three objectives do not change.

5.5 Multi-Domain Image Translation with Multi-Path Consistency

In previous two sections, we introduce image translation in two domains. Lin et al. [17] consider multi-domain image translation. In addition to the dual reconstruction error and the discrimination loss, they introduce a new kind of loss, the multi-path consistency loss, which evaluates the differences between the direct translation from a source domain to a target domain and an indirect translation from the source domain to an auxiliary domain and then to the target domain, to regularize training.

Let's consider a three-domain image translation problem as shown in Fig. 5.8, which targets at changing the hair color of an input image in a domain to another color. Ideally, the direct translation (i.e., one-hop translation) from brown hair to blond should be consistent with the indirect translation (i.e., two-hop translation) from brown to black to blond. However, such an important property is ignored in previous literature. As shown in Fig. 5.8a, without multi-path consistency regularization, the direct translation through the one-hop path and the indirect translation from the two-hop path are not consist in terms of hair color. To keep the two generated images consistent, Lin et al. [17] propose to explicitly leverage multi-path consistency to regularize the model training, which requires that the differences between the direct translation from a source to a target domain and the indirect translation from the source to an auxiliary to the target domain should be minimized. For example, in Fig. 5.8, the L_1-norm loss of the two translated blond hair images should be minimized. After applying this constraint, as shown in Fig. 5.8b, the direct and indirect translations are much similar, leading to consistent translation results.

Multi-path consistency regularization can be generally applied in image to image translation tasks. For multi-domain ($\geq$3) translations, during each training iteration, we can randomly select three domains, apply the multi-path consistency loss to

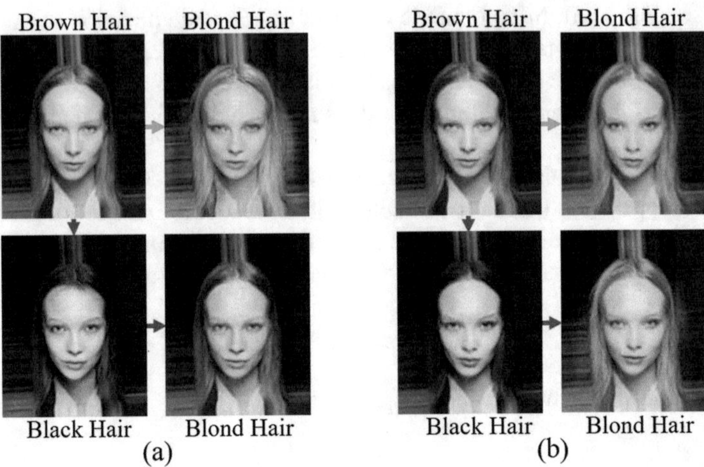

Fig. 5.8 Motivation of multi-path consistency. (**a**) Inconsistent translations. (**b**) Consistent translations. Image reproduced with permission from Lin et al. [17]

each translation task, and eventually obtain models that can generate better images. Figure 5.9 shows the data flow and training objectives for three domains: i, j, k, in which domain j is the auxiliary domain and six translators are involved.

Note that the multi-path consistency can also be applied for two-domain image translations, in which we need to introduce a third auxiliary domain to help establish the multi-path consistency.

5.6 Beyond Image Translation

Dual learning has been studied in many image and vision related problems other than image translation.

5.6.1 *Face Related Tasks*

The principle of dual reconstruction has been leveraged in many face related tasks.

Face completion aims at filling the missing or occluded regions with semantically consistent contents in face images. Li et al. [12, 15] study face completion under structured occlusions, and treat face completion and corruption as disentangling and fusing processes of clean faces and occlusions. The two processes are unified into a dual learning framework along with an adversarial strategy to learn from unpaired data.

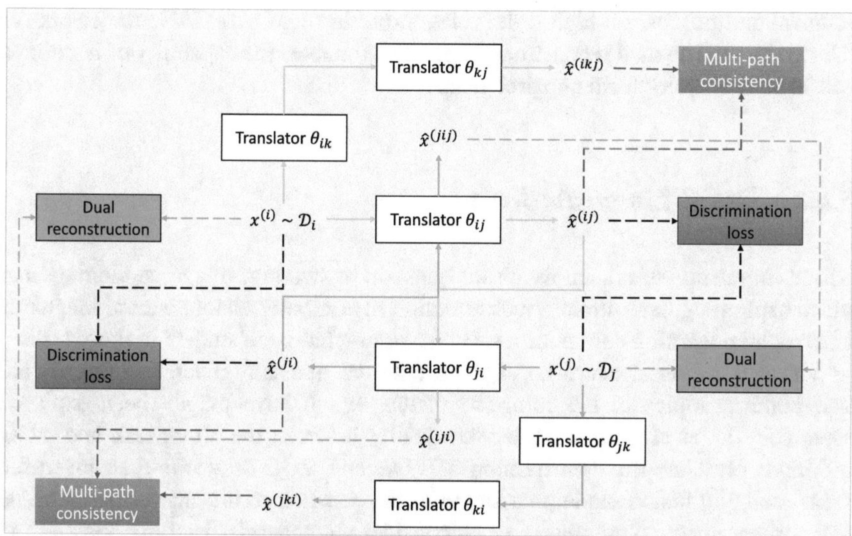

Fig. 5.9 Data flow and training objectives of multi-path consistency regularized image to image translation. Yellow solid arrows make up translation paths starting from domain i, green solid arrows make up translation paths starting from domain j, black dashed arrows make up discrimination loss, blue dashed arrows make up dual reconstruction loss, and red dashed arrows make up multi-path consistence loss

Facial action unit (AU) recognition requires fully AU-annotated facial images. Compared to facial expression labeling, annotating facial action units is time-consuming, expensive, and error-prone. Inspired by dual learning, Wang and Feng [25] propose a weakly supervised dual learning mechanism to train facial action unit classifiers from images with expression annotations. Action unit recognition from facial images is treated as the primal and main task, and face synthesis from given action units as the dual and auxiliary task. For a face image, it is first converted to action units using the primal model, and then another face image is synthesized and reconstructed from the action units using the dual model. Two kinds of rewards are used for model training: the consistency between the recognized action units and the expression label of the face, and the dual reconstruction error. By optimizing the dual tasks simultaneously, their intrinsic connections as well as domain knowledge about expressions and action units are successfully leveraged to facilitate the learning of action unit classifiers from expression-annotated images. The weakly supervised dual learning mechanism is further extended to semi-supervised dual learning settings with partially AU-annotated images.

Privacy protection becomes a critical issue as AI and machine learning has become more and more powerful. Protecting person's face photos from being misused has attracted much research attention due to the rapid development of ubiquitous face sensors. MeshFace provides a simple yet inexpensive way to protect facial photos. Li et al. [14] leverage the duality between MeshFace generation and removal and proposes a high-order relation-preserving CycleGAN framework for the two tasks. The dual reconstruction principle enables the capability of learning a MeshFace generator from unpaired data.

5.6.2 Visual-Linguistic Tasks

Visual-linguistic tasks such as visual question answering, image captioning, and video captioning have attract much attention in computer vision, natural language, and machine learning communities. Dual learning has been studied in those tasks.

Visual question answering (VQA) and visual question generation (VQG) are two trending topics in the computer vision, which have mainly been explored separately. Li et al. [13] consider the duality between the VQA task and VGQ task and formulate the dual training of VQA and VQG as learning an invertible cross-modality fusion model that can infer the question or the answer based on its and a given image. Note that Li et al. consider the supervised setting and add the dual reconstruction constraint in representations of questions and answers instead of text sequences of questions and answers. Xu et al. [31] also study visual question answering and visual question generation. They focus on VQG and leverage VQA to boost VQG through the closed loop of dual learning.

Image captioning aims to automatically generate text descriptions for a given image. Obtaining rich annotated data for images is a time-consuming and more expensive than labeling image categories (for image classification). Zhao et al. [37] propose a dual learning mechanism to address this challenge, in which the primal task is to generate image descriptions in text and the dual task is to generate plausible images from text descriptions. They consider cross-domain image captioning and their proposed method can be trained in both the semi-supervised and unsupervised adaptation settings. A similar approach is also proposed in [32].

Video captioning, which aims to describe a video with natural language automatically, has received much attention in both the computer vision and natural language processing communities. Wang et al. [26] introduce the sentence-to-video task as the dual task to boost video captioning. They propose a reconstruction network with an encoder-decoder-reconstructor architecture to exploit the duality between the forward and backward flows. In the forward flow, the encoder encodes each video frame into a semantic representation, from which the decoder generates a sentence description. In the backward flow, the reconstructor reproduces the feature sequence of the original video based on the hidden state sequence of the decoder. In addition to the standard supervised objective, based on the dual reconstruction principle, the training is enhanced with the minimization of the differences between original and reproduced video features. A refined model is proposed to leverage both global and local structures to further improve the reconstruction of the video representation [36].

Album storytelling, which is related to but different from image and video captioning, aims to generate text descriptions for a set of visually related or uncorrelated images. Similar to [26, 36], dual learning is employed to boost the performance of album storytelling by reconstructing the album representations from the hidden representations of the decoder [27].

Dual learning has also been studied in other visual-linguistic tasks, such as text-image retrieval and matching [3, 18].

5.6.3 Other Image Related Tasks

Luo et al. [19] study dual learning for semantic image segmentation, which aims to assign a semantic label such as 'dog', 'flower', and 'cat' to each pixel in an image. Labeled training data for semantic image segmentation is usually limited, because the per-pixel label maps are difficult and expensive to obtain. To reduce labeling efforts, a natural solution is to collect additional images from Internet that are associated with image-level tags. Different from previous works that treat label maps and tags as independent supervisions, Luo et al. present a novel learning setting, dual image segmentation (DIS), which jointly solve two complementary learning problems based on the principle of dual reconstruction: One problem is to predict label maps and tags from images, and the other is to reconstruct the images using the predicted label maps.

Estimating multi-type cardiac indices from cardiac magnetic resonance imaging (MRI) and computed tomography (CT) images attracts much attention recently because of its clinical potential for comprehensive function assessment. Yu et al. [34] introduce a dual task, cardiac image generation from multi-type cardiac indices, and leverage the structural duality between the two tasks based on the dual reconstruction principle to boost the primal task of multi-type cardiac indices estimation.

Dual learning has also been studied in zero-shot and few-shot image classification [7, 35], to improve the classification accuracy with a few or even zero labeled images.

References

1. Arjovsky, M., Chintala, S., & Bottou, L. (2017). Wasserstein generative adversarial networks. In *International Conference on Machine Learning* (pp. 214–223).
2. Choi, Y., Choi, M., Kim, M., Ha, J.-W., Kim, S., & Choo, J. (2018). StarGAN: Unified generative adversarial networks for multi-domain image-to-image translation. In *Proceedings of the IEEE Conference on Computer Vision and Pattern Recognition* (pp. 8789–8797).
3. Cornia, M., Baraldi, L., Tavakoli, H. R., & Cucchiara, R. (2020). A unified cycle-consistent neural model for text and image retrieval. *Multimedia Tools and Applications*, 1–25.
4. Goodfellow, I. (2016). Nips 2016 tutorial: Generative adversarial networks. Preprint. arXiv:1701.00160.

5. Goodfellow, I., Pouget-Abadie, J., Mirza, M., Xu, B., Warde-Farley, D., Ozair, S., et al. (2014). Generative adversarial nets. In *Advances in Neural Information Processing Systems* (pp. 2672–2680).
6. Gulrajani, I., Ahmed, F., Arjovsky, M., Dumoulin, V., & Courville, A. C. (2017). Improved training of Wasserstein GANs. In *Advances in Neural Information Processing Systems* (pp. 5767–5777).
7. Huang, H., Wang, C., Yu, P. S., & Wang, C.-D. (2019). Generative dual adversarial network for generalized zero-shot learning. In *Proceedings of the IEEE Conference on Computer Vision and Pattern Recognition* (pp. 801–810).
8. Isola, P., Zhu, J.-Y., Zhou, T., & Efros, A. A. (2017). Image-to-image translation with conditional adversarial networks. In *Proceedings of the IEEE Conference on Computer Vision and Pattern Recognition* (pp. 1125–1134).
9. Kim, T., Cha, M., Kim, H., Lee, J. K., & Kim, J. (2017). Learning to discover cross-domain relations with generative adversarial networks. In *Proceedings of the 34th International Conference on Machine Learning* (Vol. 70, pp. 1857–1865). JMLR.org.
10. Larsen, A. B. L., Sønderby, S. K., Larochelle, H., & Winther, O. (2016). Autoencoding beyond pixels using a learned similarity metric. In *International Conference on Machine Learning* (pp. 1558–1566).
11. Li, C., & Wand, M. (2016). Precomputed real-time texture synthesis with Markovian generative adversarial networks. In *European Conference on Computer Vision* (pp. 702–716). New York: Springer.
12. Li, Z., Hu, Y., & He, R. (2017). Learning disentangling and fusing networks for face completion under structured occlusions. Preprint. arXiv:1712.04646.
13. Li, Y., Duan, N., Zhou, B., Chu, X., Ouyang, W., Wang, X., et al. (2018). Visual question generation as dual task of visual question answering. In *Proceedings of the IEEE Conference on Computer Vision and Pattern Recognition* (pp. 6116–6124).
14. Li, Z., Hu, Y., Zhang, M., Xu, M., & He, R. (2018). Protecting your faces: Meshfaces generation and removal via high-order relation-preserving cyclegan. In *2018 International Conference on Biometrics (ICB)* (pp. 61–68). New York: IEEE.
15. Li, Z., Hu, Y., He, R., & Sun, Z. (2020). Learning disentangling and fusing networks for face completion under structured occlusions. *Pattern Recognition, 99*, 107073.
16. Lin, J., Xia, Y., Qin, T., Chen, Z., & Liu, T.-Y. (2018). Conditional image-to-image translation. In *Proceedings of the IEEE Conference on Computer Vision and Pattern Recognition* (pp. 5524–5532).
17. Lin, J., Xia, Y., Wang, Y., Qin, T., & Chen, Z. (2019). Image-to-image translation with multi-path consistency regularization. In *Proceedings of the Twenty-Eighth International Joint Conference on Artificial Intelligence* (pp. 2980–2986).
18. Liu, Y., Guo, Y., Liu, L., Bakker, E. M., & Lew, M. S. (2019). Cyclematch: A cycle-consistent embedding network for image-text matching. *Pattern Recognition, 93*, 365–379.
19. Luo, P., Wang, G., Lin, L., & Wang, X. (2017). Deep dual learning for semantic image segmentation. In *Proceedings of the IEEE International Conference on Computer Vision* (pp. 2718–2726).
20. Mao, X., Li, Q., Xie, H., Lau, R. Y. K., Wang, Z., & Smolley, S. P. (2017). Least squares generative adversarial networks. In *Proceedings of the IEEE International Conference on Computer Vision* (pp. 2794–2802).
21. Metz, L., Poole, B., Pfau, D., & Sohl-Dickstein, J. (2017). Unrolled generative adversarial networks. In *5th International Conference on Learning Representations.*
22. Radford, A., Metz, L., & Chintala, S. (2015). Unsupervised representation learning with deep convolutional generative adversarial networks. Preprint. arXiv:1511.06434.
23. Ronneberger, O., Fischer, P., & Brox, T. (2015). U-net: Convolutional networks for biomedical image segmentation. In *International Conference on Medical Image Computing and Computer-Assisted Intervention* (pp. 234–241). Cham: Springer.

24. Salimans, T., Goodfellow, I., Zaremba, W., Cheung, V., Radford, A., & Chen, X. (2016). Improved techniques for training GANs. In *Advances in Neural Information Processing Systems* (pp. 2234–2242).
25. Wang, S., & Peng, G. (2019). Weakly supervised dual learning for facial action unit recognition. *IEEE Transactions on Multimedia, 21*(12), 3218–3230.
26. Wang, B., Ma, L., Zhang, W., & Liu, W. (2018). Reconstruction network for video captioning. In *Proceedings of the IEEE Conference on Computer Vision and Pattern Recognition* (pp. 7622–7631).
27. Wang, B., Ma, L., Zhang, W., Jiang, W., & Zhang, F. (2019). Hierarchical photo-scene encoder for album storytelling. In *Proceedings of the AAAI Conference on Artificial Intelligence* (Vol. 33, pp. 8909–8916).
28. Wang, Y., Xia, Y., He, T., Tian, F., Qin, T., Zhai, C. X., et al. (2019). Multi-agent dual learning. In *7th International Conference on Learning Representations, ICLR 2019*.
29. Wang, Z., She, Q., & Ward, T. E. (2019). Generative adversarial networks in computer vision: A survey and taxonomy. Preprint. arXiv:1906.01529.
30. Wu, L., Xia, Y., Tian, F., Zhao, L., Qin, T., Lai, J., et al. (2018). Adversarial neural machine translation. In *Asian Conference on Machine Learning* (pp. 534–549).
31. Xu, X., Song, J., Lu, H., He, L., Yang, Y., & Shen, F. (2018). Dual learning for visual question generation. *2018 IEEE International Conference on Multimedia and Expo (ICME)* (pp. 1–6).
32. Yang, M., Zhao, W., Xu, W., Feng, Y., Zhao, Z., Chen, X., et al. (2018). Multitask learning for cross-domain image captioning. *IEEE Transactions on Multimedia, 21*(4), 1047–1061.
33. Yi, Z., Zhang, H., Tan, P., & Gong, M. (2017). Dualgan: Unsupervised dual learning for image-to-image translation. In *Proceedings of the IEEE International Conference on Computer Vision* (pp. 2849–2857).
34. Yu, C., Gao, Z., Zhang, W., Yang, G., Zhao, S., Zhang, H., et al. (2020). Multitask learning for estimating multitype cardiac indices in MRI and CT based on adversarial reverse mapping. *IEEE Transactions on Neural Networks and Learning Systems, 99*, 1–14.
35. Zhang, C., Lyu, X., & Tang, Z. (2019). TGG: Transferable graph generation for zero-shot and few-shot learning. In *Proceedings of the 27th ACM International Conference on Multimedia* (pp. 1641–1649).
36. Zhang, W., Wang, B., Ma, L., & Liu, W. (2019). Reconstruct and represent video contents for captioning via reinforcement learning. *IEEE Transactions on Pattern Analysis and Machine Intelligence*.
37. Zhao, W., Xu, W., Yang, M., Ye, J., Zhao, Z., Feng, Y., et al. (2017). Dual learning for cross-domain image captioning. In *Proceedings of the 2017 ACM on Conference on Information and Knowledge Management* (pp. 29–38).
38. Zhou, T., Krahenbuhl, P., Aubry, M., Huang, Q., & Efros, A. A. (2016). Learning dense correspondence via 3d-guided cycle consistency. In *Proceedings of the IEEE Conference on Computer Vision and Pattern Recognition* (pp. 117–126).
39. Zhu, J.-Y., Park, T., Isola, P., & Efros, A. A. (2017). Unpaired image-to-image translation using cycle-consistent adversarial networks. In *Proceedings of the IEEE International Conference on Computer Vision* (pp. 2223–2232).

Chapter 6
Dual Learning for Speech Processing and Beyond

6.1 Neural Speech Synthesis and Recognition

Speech synthesis (or text to speech, TTS) and automatic speech recognition (ASR) are two important tasks in speech processing and have long been hot research topics in the field of artificial intelligence.

Speech synthesis systems have been evolved for multiple stages, from early formant synthesis, articulatory speech synthesis, concatenative approaches, to statistical parametric approaches, to today's deep learning based neural approaches. In recent years, neural networks based deep learning has become the dominant approach for speech synthesis and recognition.

TTS and ASR are typical sequence-to-sequence learning problems. Recent successes of deep learning methods have pushed TTS and ASR into end-to-end learning, where both tasks can be modeled in an encoder-decoder framework with attention mechanism.[1] CNN/RNN based models are widely used in TTS and ASR [3, 7, 19, 20, 24, 30].

As deep neural models are all data-hungry, this brings challenges to neural speech synthesis and recognition for many languages that are scarce of paired speech and text data. Therefore, many techniques for low-resource and zero-resource speech synthesis and recognition have been proposed recently, including unsupervised speech recognition [4, 5, 17, 36], low-resource ASR [8, 9, 40], and TTS with minimal speaker data [1, 6, 12, 31].

As aforementioned in Chap. 1, speech synthesis and recognition are naturally in dual form. It is obvious that structure duality can help speech synthesis and recognition models to learn from unlabeled data (i.e., unpaired speech and text data).

---

[1]Although some previous works adopt a feed-forward network [35, 37] and achieved promising results on ASR, the encoder-decoder framework becomes more popular and successful in past years.

T. Qin, *Dual Learning*, https://doi.org/10.1007/978-981-15-8884-6_6

We introduce several recent works [21, 22, 28, 29, 34] on dual learning for speech synthesis and recognition.

6.2 Speech Chain with Dual Learning

The speech chain was introduced to describe the basic mechanism of speech communication in which a spoken message travels from the brain of a speaker to the brain of a listener. It consists of the speech production process, in which the speaker produces speech sound, and a speech perception process, in which the listener hears and perceives what was said. In speech communication, the hearing is not only critical for the listener but also for the speaker. By simultaneously listening and speaking, the speaker can monitor her speech (e.g., volume, articulation, and overall comprehensibility) and better plan what and how she will say next. Children who lose their hearing often have difficulty to produce clear speech due to inability to monitor their own speech.

Inspired by the importance of close-loop feedback in the human speech chain, Tjandra et al. [28] are among the first to conduct joint training for speech synthesis and recognition and learn from unpaired data by leveraging the close-loop feedback of dual reconstruction. They call their proposed system the machine speech chain, the overall architecture of which is shown in Fig. 6.1.

As shown in the figure, the machine speech chain consists a TTS module, which converts a text sequence to a speech sequence, an ASR module, which converts a speech sequence to a text sequence, and a reconstruction loop. Its key idea is the joint training of the ASR and TTS models based on the dual reconstruction principle, with both paired data and unpaired data.

Before introducing the training objectives, we first give some notations. Let $\mathcal{D}$ denote the set of paired speech and text sequences, $\mathcal{X}$ the set of unpaired speech sequences, and $\mathcal{Y}$ the set of unpaired text sequences. Let θ denote the parameters of the TTS model and ϕ denote the parameters of the ASR model.

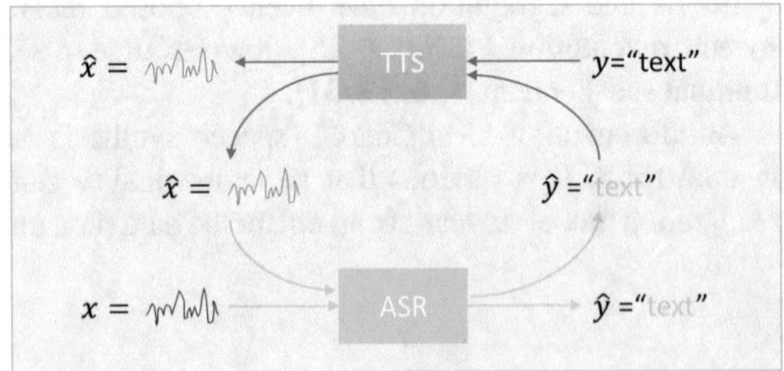

Fig. 6.1 Overall architecture of machine speech chain

For paired data $\mathcal{D}$, the training of the TTS model is to minimize the following mean square error

$$l(\theta; \mathcal{D}) = \frac{1}{|\mathcal{D}|} \sum_{(x,y)\in\mathcal{D}} (x - f(y; \theta))^2, \tag{6.1}$$

where (x, y) is a pair of speech text sequences, $f(\cdot; \theta)$ is the TTS model with parameter θ. Note that a speech sequence is usually converted into a sequence of mel-spectrograms, and the mean square error is calculated on mel-spectrograms. Tjandra et al. [28] adopt a sequence-to-sequence TTS model, the core network architecture of which is based on Tacotron [30].

The training of the ASR model is to minimize the following negative log likelihood

$$l(\phi; \mathcal{D}) = -\frac{1}{|\mathcal{D}|} \sum_{x,y)\in\mathcal{D}} \log P(y|x; \phi), \tag{6.2}$$

where $P(y|x; \phi)$ is the probability of the ASR model with parameter ϕ generating text sequence y from the speech sequence x. Tjandra et al. [28] adopt an attention-based encoder-decoder model for ASR, which is also a sequence-to-sequence model as introduced in Sect. 3.3.2.

For an unpaired speech sequence x in $\mathcal{X}$, the dual reconstruction loop first converts it to a text sequence $\hat{y}(x)$ using the ASR model:

$$\hat{y}(x) = \arg\max_y P(y|x; \phi),$$

and then reconstructs the original x from $\hat{y}(x)$ using the TTS model θ. That is, the training objective is to minimize the following mean square error.

$$l(\theta; \mathcal{X}) = \frac{1}{|\mathcal{X}|} \sum_{x\in\mathcal{X}} (x - f(\hat{y}(x); \theta))^2 \tag{6.3}$$

Similarly, for an unpaired text sequence y in $\mathcal{Y}$, the dual reconstruction loop first converts it to a speech sequence $\hat{x}(y)$ using the TTS model:

$$\hat{x}(y) = f(y; \theta),$$

and then reconstructs the original y from $\hat{x}(y)$ using the ASR model ϕ. That is, the training objective is to minimize the following negative log likelihood.

$$l(\phi; \mathcal{Y}) = -\frac{1}{|\mathcal{Y}|} \sum_{y\in\mathcal{Y}} \log P(y|\hat{x}(y); \phi) \tag{6.4}$$

The overall training procedure of machine speech chain is shown in Algorithm 1. Experiments [28] on both single-speaker synthetic speech and multi-speaker natural speech demonstrate that the speech chain algorithm can effectively improve both the TTS and ASR models using unpaired data through the dual reconstruction loop.

A limitation of the speech chain algorithm is that it is unable to handle unseen speakers. Tjandra et al. [29] extend the algorithm by integrating a speaker recognition model inside the dual loop and enable the capability of TTS to handle unseen speakers by implementing one-shot speaker adaptation. The one-shot speaker adaptation enables TTS to mimic voice characteristics from one speaker to another with only one speech sample, even from a text sequence without any speaker information. Furthermore, ASR in the dual reconstruction loop also benefits from the ability to learn an arbitrary speaker's characteristics from generated speech sequences, resulting in significant improvement of recognition accuracy [29].

Algorithm 1 The speech chain algorithm

Require: Paired data $\mathcal{D}$, unpaired speech data $\mathcal{X}$, unpaired text data $\mathcal{Y}$.

Randomly initialize the TTS model θ and the ASR model ϕ.

repeat

  Sample a minbatch of data D from $\mathcal{D}$.

  Compute the loss $\ell(\theta; D)$ according to Eq. (6.1) over the data D.

  Compute the loss $\ell(\phi; D)$ according to Eq. (6.2) over the data D.

  Sample a minbatch of data X from $\mathcal{X}$.

  Compute the loss $\ell(\theta; X)$ according to Eq. (6.3) over the data X.

  Sample a minbatch of data Y from $\mathcal{Y}$.

  Compute the loss $\ell(\phi; Y)$ according to Eq. (6.4) over the data Y.

  Compute the gradient of θ and ϕ:

$$\Delta\theta = \nabla_\theta[\ell(\theta; D) + \ell(\theta; X)],$$

$$\Delta\phi = \nabla_\phi[\ell(\phi; D) + \ell(\phi; Y)].$$

  Update models with any optimizers opt_1 and opt_1:

$$\theta \leftarrow opt_1(\theta, \Delta\theta), \quad \phi \leftarrow opt_1(\phi, \Delta\phi).$$

until the convergence of θ and ϕ.

6.3 Dual Learning for Low-Resource Speech Processing

Ren et al. [22] study speech synthesis and recognition in the low-resource setting, in which only a small number of paired speech text sequences are available and most training data are unpaired speech sequences and text sequences. Different from the speech chain algorithm introduced in previous section, in addition to dual learning, Ren et al. [22] adopt the idea of denoising auto-encoding inspired by unsupervised machine translation, and introduce bidirectional sequence modeling to handle error propagation in sequence learning.

In sequence-to-sequence learning, error propagation [2, 25] refers to the problem that if an token is mistakenly predicted during inference, the error will be propagated and the future tokens conditioned on this one will be impacted. This will result in that the right part of a generated sequence is worse than the left part. The speech and text sequences[2] are usually longer than the sequence in other NLP tasks such as neural machine translation, and thus TTS and ASR suffer more from error propagation, especially in the low-resource setting.

In order to solve the above problem, Ren at al. [22] leverage the bidirectional sequence modeling to generate speech and text sequence in both left-to-right and right-to-left directions. In this way, the right part of the sequence that is always of low quality in the original dual transformation process can be generated in the right-to-left direction with good quality. Consequently, the dual task that relies on the generated data for training will benefit from the improved quality on the right part of the sequence and result in higher transformation accuracy than the original left-to-right generation. At the same time, bidirectional sequence modeling can also act as an effect of data augmentation that leverages the data in both directions, which is helpful especially when few paired data are available in the low-resource setting.

In the following sub section, we introduce the training objectives for denoising auto-encoding and dual reconstruction.

6.3.1 *Denoising Auto-Encoding with Bidirectional Sequence Modeling*

Similar to the denoising auto-encoding in unsupervised neural machine translation (Sect. 4.4), Ren et al. reconstruct the original clean input of speech and text sequences from the corrupted version, as shown in Fig. 6.2a, b. The loss function ℓ_{dae} of the denoising auto-encoding on speech and text data is as follows:

$$\ell_{dae} = \sum_{x \in \mathcal{S}} \ell_{sp}(x|c(x); \theta_{en}^{sp}, \theta_{de}^{sp}) + \sum_{y \in \mathcal{T}} \ell_{tx}(y|c(y); \theta_{en}^{tx}, \theta_{de}^{tx}), \tag{6.5}$$

where $\mathcal{S}$ and $\mathcal{T}$ denote the set of unpaired sequences in the speech and text domain, $\theta_{en}^{sp}, \theta_{de}^{sp}, \theta_{en}^{tx}$ and θ_{de}^{tx} denote the model parameters of the speech encoder, the speech decoder, the text encoder, and the text decoder respectively, $c()$ is a corruption operator that randomly masks some elements with zero vectors, or swaps the elements in a certain window of the speech and text sequences [15]. ℓ_{sp} and ℓ_{tx}

[2] A speech sequence is usually converted into mel-spectrograms that contain thousands of frames, while a text sequence is usually converted into phoneme sequence and is longer than the original word or sub-word sequence.

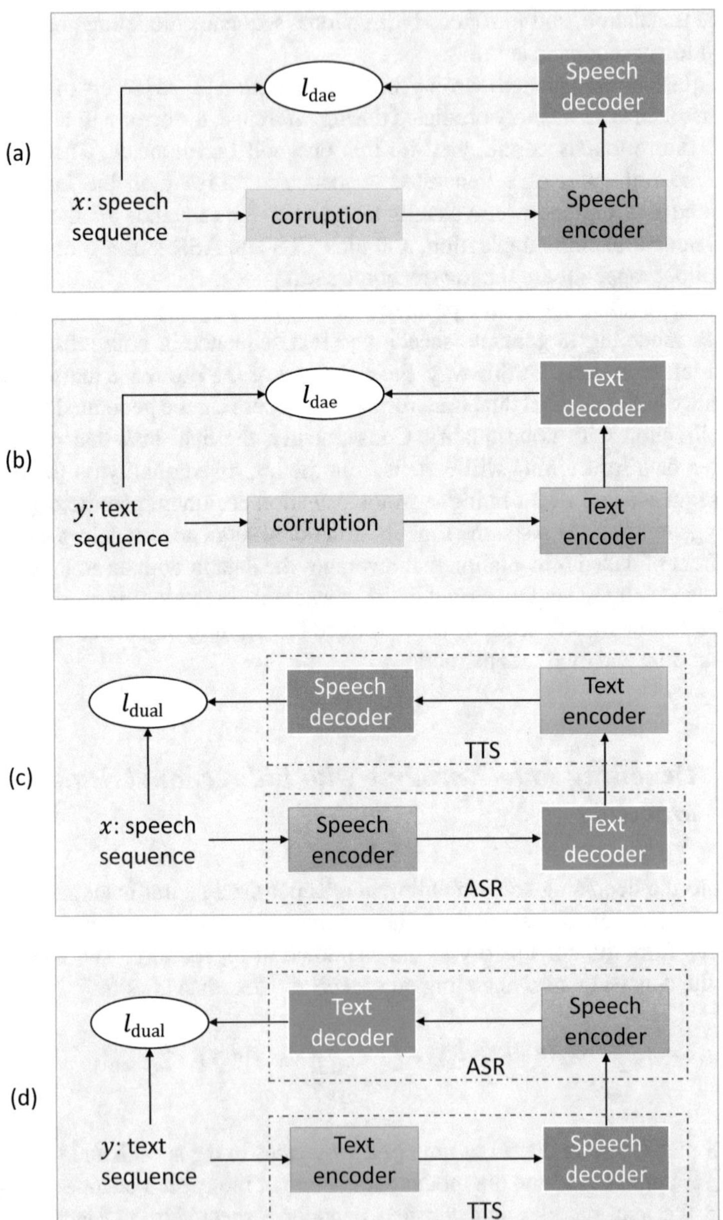

Fig. 6.2 Training flow of low-resource TTS and ASR in [22]. (**a**) Denoising auto-encoding for speech. (**b**) Denoising auto-encoding for text. (**c**) Dual reconstruction for speech. (**d**) Dual reconstruction for text. Note that the corruption operation only be used in denoising auto-encoding

denote the loss for speech and text sequence respectively. In general, we have

$$\ell_{sp}(y|x;\theta_{en}^{sp},\theta_{de}^{sp}) = \text{MSE}(y, f(x;\theta_{en}^{sp},\theta_{de}^{sp})),$$

$$\ell_{tx}(y|x;\theta_{en}^{tx},\theta_{de}^{tx}) = -\Sigma \log P(y|x;\theta_{en}^{tx},\theta_{de}^{tx})),$$

where MSE denotes the mean squared errors for speech.

To incorporate bidirectional sequence modeling, we need to re-formulate the objective of denoising auto-encoding and dual reconstruction:

$$\begin{aligned}
\ell_{\overrightarrow{dae}} &= \sum_{x\in\mathcal{S}} \ell_{sp}(\overrightarrow{x}|c(\overrightarrow{x});\theta_{en}^{sp},\theta_{de}^{sp}) + \sum_{y\in\mathcal{T}} \ell_{tx}(\overrightarrow{y}|c(\overrightarrow{y});\theta_{en}^{tx},\theta_{de}^{tx}),\\
\ell_{\overleftarrow{dae}} &= \sum_{x\in\mathcal{S}} \ell_{sp}(\overleftarrow{x}|c(\overleftarrow{x});\theta_{en}^{sp},\theta_{de}^{sp}) + \sum_{y\in\mathcal{T}} \ell_{tx}(\overleftarrow{y}|c(\overleftarrow{y});\theta_{en}^{tx},\theta_{de}^{tx}),
\end{aligned} \tag{6.6}$$

where $\overrightarrow{x}$ is the left-to-right version of sequence x, i.e., ordering tokens in x in the left-to-right manner, and $\overleftarrow{x}$ is the right-to-left version of sequence x. That is, we reconstruct the corrupted speech and text sequences in both left-to-right and right-to-left directions. The model parameters are shared while modeling the sequence in two directions.

6.3.2 *Dual Reconstruction with Bidirectional Sequence Modeling*

Ren et al. follow the dual reconstruction principle to leverage the dual nature of TTS and ASR and learn from unlabeled data for TTS and ASR. As shown in Fig. 6.2c, a speech sequence x is first transformed into a text sequence $\hat{y}$ using the ASR model, and then the TTS model is trained to reconstruct the original speech sequence x from the generated text sequence $\hat{y}$, i.e., the TTS model is trained on the pseudo pair $(\hat{y}, x)$. Similarly, in Fig. 6.2d, we train the ASR model on the pseudo pair $(\hat{x}, y)$ generated by the TTS model from a text sequence y.

The loss ℓ_{dual} for dual reconstruction consists of the following two parts:

$$\ell_{dual} = \sum_{x\in\mathcal{S}} \ell_{sp}(x|\hat{y};\theta_{en}^{tx},\theta_{de}^{sp}) + \sum_{y\in\mathcal{T}} \ell_{tx}(y|\hat{x};\theta_{en}^{sp},\theta_{de}^{tx}), \tag{6.7}$$

where $\hat{y} = \arg\max P(y|x;\theta_{en}^{sp},\theta_{de}^{tx})$ and $\hat{x} = f(y;\theta_{en}^{tx},\theta_{de}^{sp})$ denote the text and speech sequence generated from speech x and text y respectively. During model training, dual reconstruction is running on the fly, where the TTS model leverages the newest text sequence generated by the ASR model for training, and vice versa, to ensure the accuracy of TTS and ASR can gradually improve.

With bidirectional modeling, the dual reconstruction loss is rewritten as follows:

$$
\begin{aligned}
\ell_{\overrightarrow{dual}} = & \sum_{x \in \mathcal{S}} \ell_{sp}(\overrightarrow{x} \mid \hat{\overrightarrow{y}}; \theta_{en}^{tx}, \theta_{de}^{sp}) + \sum_{x \in \mathcal{S}} \ell_{sp}(\overrightarrow{x} \mid r(\hat{\overleftarrow{y}}); \theta_{en}^{tx}, \theta_{de}^{sp}) \\
& + \sum_{y \in \mathcal{T}} \ell_{tx}(\overrightarrow{y} \mid \hat{\overrightarrow{x}}; \theta_{en}^{sp}, \theta_{de}^{tx}) + \sum_{y \in \mathcal{T}} \ell_{tx}(\overrightarrow{y} \mid r(\hat{\overleftarrow{x}}); \theta_{en}^{sp}, \theta_{de}^{tx}), \\
\ell_{\overleftarrow{dual}} = & \sum_{x \in \mathcal{S}} \ell_{sp}(\overleftarrow{x} \mid \hat{\overleftarrow{y}}; \theta_{en}^{tx}, \theta_{de}^{sp}) + \sum_{x \in \mathcal{S}} \ell_{sp}(\overleftarrow{x} \mid r(\hat{\overrightarrow{y}}); \theta_{en}^{tx}, \theta_{de}^{sp}) \\
& + \sum_{y \in \mathcal{T}} \ell_{tx}(\overleftarrow{y} \mid \hat{\overleftarrow{x}}; \theta_{en}^{sp}, \theta_{de}^{tx}) + \sum_{y \in \mathcal{T}} \ell_{tx}(\overleftarrow{y} \mid r(\hat{\overrightarrow{x}}); \theta_{en}^{sp}, \theta_{de}^{tx}),
\end{aligned} \tag{6.8}
$$

where $r(\cdot)$ is the reverse function that reverses the sequence from left-to-right to right-to-left or the other way around,

$$
\begin{aligned}
\hat{\overrightarrow{y}} &= \arg\max P(\overrightarrow{y} \mid \overrightarrow{x}; \theta_{en}^{sp}, \theta_{de}^{tx}), \\
\hat{\overleftarrow{y}} &= \arg\max P(\overleftarrow{y} \mid \overleftarrow{x}; \theta_{en}^{sp}, \theta_{de}^{tx}), \\
\hat{\overrightarrow{x}} &= f(\overrightarrow{y}; \theta_{en}^{tx}, \theta_{de}^{sp}), \\
\hat{\overleftarrow{x}} &= f(\overleftarrow{y}; \theta_{en}^{tx}, \theta_{de}^{sp})
\end{aligned}
$$

denoting the sequence generated from x and y in left-to-right and right-to-left directions respectively. The loss term $\ell_{sp}(\overrightarrow{x} \mid r(\hat{\overleftarrow{y}}); \theta_{en}^{tx}, \theta_{de}^{sp})$ and $\ell_{tx}(\overrightarrow{y} \mid r(\hat{\overleftarrow{x}}); \theta_{en}^{sp}, \theta_{de}^{tx})$ in $\ell_{\overrightarrow{dual}}$ can help the model to better learn on the right part of the sequence, which is usually of poor quality due to error propagation. Similar loss terms can be found in $\ell_{\overleftarrow{dual}}$.

6.3.3 Model Training

As shown in Eqs. (6.6) and (6.8), the left-to-right generation and right-to-left generation share the same model, i.e., we train one model that bidirectionally generates sequences. In order to give the model a sense of which direction the sequence will be generated, unlike the conventional decoder using a zero vector as the start token for training and inference, Ren et al. use two learnable direction vectors to indicate the generation directions, one for left to right generation and the other for right to left generation. In total, there are four direction vectors in total, two for speech generation and the other two for text generation.

In addition to unlabeled data, similar to DualNMT in Sect. 4.3, Ren et al. also leverage a few paired data for bidirectional training, with the following bidirectional supervised loss:

$$\begin{aligned}\ell_{\overrightarrow{sup}} &= \sum_{(x,y)\in\mathcal{D}} \ell_{sp}(\overrightarrow{x} \mid \overrightarrow{y}; \theta_{en}^{tx}, \theta_{de}^{sp}) + \sum_{(x,y)\in\mathcal{D}} \ell_{tx}(\overrightarrow{y} \mid \overrightarrow{x}; \theta_{en}^{sp}, \theta_{de}^{tx}),\\ \ell_{\overleftarrow{sup}} &= \sum_{(x,y)\in\mathcal{D}} \ell_{sp}(\overleftarrow{x} \mid \overleftarrow{y}; \theta_{en}^{tx}, \theta_{de}^{sp}) + \sum_{(x,y)\in\mathcal{D}} \ell_{tx}(\overleftarrow{y} \mid \overleftarrow{x}; \theta_{en}^{sp}, \theta_{de}^{tx}),\end{aligned} \tag{6.9}$$

where $\mathcal{D}$ denotes the paired speech and text sequences.

The total loss of the method proposed in [22] is as follows:

$$\ell = \ell_{\overrightarrow{dae}} + \ell_{\overleftarrow{dae}} + \ell_{\overrightarrow{dual}} + \ell_{\overleftarrow{dual}} + \ell_{\overrightarrow{sup}} + \ell_{\overleftarrow{sup}}, \tag{6.10}$$

where each individual loss is described in Eqs. (6.6), (6.8), and (6.9).

Ren et al. conducted experiments on the LJSpeech dataset [11] by leveraging only 200 paired speech and text sentences and extra unpaired data for training. The results show that their method can generate intelligible voice with a word level intelligible rate of 99.84%, compared with nearly 0 intelligible rate if training on only 200 paired sentences, and achieve 2.68 MOS (mean opinion score) for TTS and 11.7% PER (phoneme error rate) for ASR, outperforming the baseline model trained on only 200 paired sentences. The speech samples generated by their method and the baseline model can be found at https://speechresearch.github.io/unsuper/.

6.4 Dual Learning for Extremely Low-Resource Speech Processing

Although Ren et al.[22] produce reasonably good artificial speech and significantly improve the accuracy of speech recognition for the low-resource setting by leveraging unpaired speech and text data through dual learning, the quality of synthesized speech and the accuracy of the obtained speech recognition model still do not meet the requirement of commercial speech services. Going beyond [22] and taking one step further, Xu et al. [34] extend dual learning to the extremely low-resource setting for speech synthesis and speech recognition and target at industrial deployment under two constraints: (1) extremely low data collection cost, and (2) high accuracy to satisfy the deployment requirement.

Industrial neural speech synthesis and recognition systems typically use several kinds of data for model training.

- Speech synthesis needs high-quality single-speaker (the target speaker) recordings that are collected in professional recording studio. To improve the pronunciation accuracy, speech synthesis also requires a pronunciation lexicon to convert the character sequence into phoneme sequence as the model input (e.g., "speech" is converted into "s p iy ch"), which is called as grapheme-to-phoneme conversion [26]. In addition, speech synthesis models use text normalization rules to convert the irregular word into the normalized type that is easier to pronounce (e.g., "Dec 6th" is converted into "December sixth").

- Speech recognition requires speech data from multiple speakers in order to generalize to unseen speakers during inference. The multi-speaker speech data in speech recognition do not need to be as high-quality as that in speech synthesis, but the data amount is usually an order of magnitude larger. The speech data for speech recognition is called multi-speaker low-quality data.[3] Optionally, a speech recognition model can first recognize a speech sequence into a phoneme sequence, and then convert it into a character sequence with the pronunciation lexicon as used in speech synthesis.
- In addition to paired speech and text data, speech synthesis and recognition systems may also leverage unpaired speech and text data for model training.

Xu et al. [34] reduce the overall data collection cost considering that different kinds of data are of different cost:

- They only use several minutes of single-speaker high-quality paired speech text data, because this kind of data is the most costly to collect and the speech data is usually recorded in professional studios.
- They use several hours of multi-speaker low-quality paired speech text data. The low-quality paired data is relatively less costly to collect compared with the above high-quality paired data. For example, one can collect speech data from the web and ask human labelers to transcribe them.
- They use dozens of hours of multi-speaker low-quality unpaired speech data. This kind of low-quality speech without paired text is relatively easy to collect from the Web, at almost zero cost.
- They do not use high-quality speech data from the target TTS speaker, which is costly to collect.
- They do not use the pronunciation lexicon but directly take characters as the input of the TTS model and the output of the ASR model, because the lexicon is costly to obtain, especially for rare languages.

To ensure the TTS quality and ASR accuracy while reducing data cost, Xu et al. [34] develop a system for joint speech synthesis and recognition, named LRSpeech, which is based on three key techniques:

1. pre-training on rich-resource languages and fine-tuning on the low-resource target language,

[3]Here low quality does not mean the transcripts are incorrect. Instead, it means that the quality of speech data is just relatively low (e.g., with background noise, incorrect pronunciations, etc.) compared with the high-quality TTS recordings.

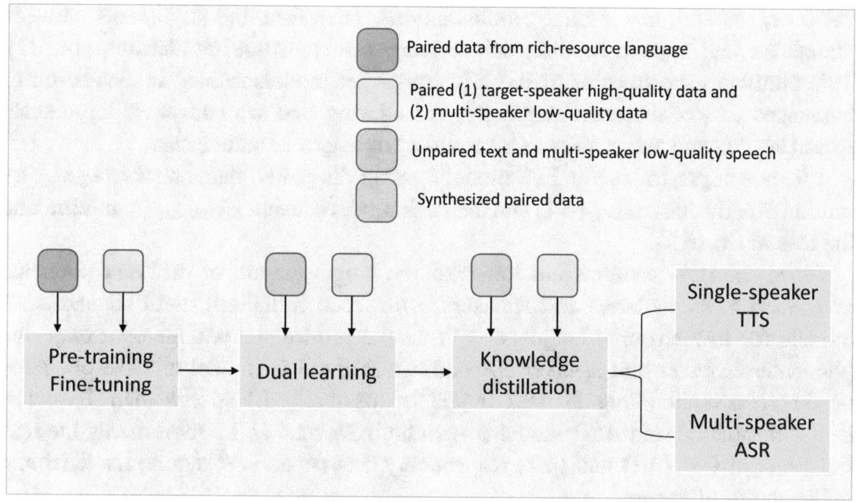

Fig. 6.3 The three-stage pipeline of LRSpeech

2. iterative accuracy boosting between speech synthesis and recognition through dual learning, and
3. knowledge distillation to further improve synthesis quality and recognition accuracy.

Specifically, the training pipeline of LRSpeech contains three stages, as shown in Fig. 6.3:

Before introducing the details of the three stages, we first define some notations. Let $\mathcal{D}$ denote a paired text and speech corpus and $(x, y) \in \mathcal{D}$ denote a speech and text sequence pair. Each element x_i in the speech sequence x represents a frame of speech, and each element y_i in the text sequence y represents a phoneme or character.

Let $\mathcal{D}_{rich\_tts}$ denote as the high-quality TTS paired data in rich-resource languages, $\mathcal{D}_{rich\_asr}$ the low-quality multi-speaker ASR paired data in rich-resource languages, $\mathcal{D}_h$ the single-speaker high-quality paired data of the target low-resource language for the target speaker, and $\mathcal{D}_l$ the multi-speaker low-quality paired data of the target low-resource language. Let $\mathcal{X}^u$ denote multi-speaker low-quality unpaired speech data and $\mathcal{Y}^u$ unpaired text data.

6.4.1 Pre-training and Fine-Tuning

Learning the alignment between the character/phoneme representations (text) and the acoustic features (speech) is the key to TTS and ASR. Although people coming from different nations may speak different languages, they share similar vocal organs and thus similar pronunciations. Therefore, the alignments between phonemes and speech in different languages share some similarities [14, 32]. This motivates the transfer of the TTS and ASR models trained in rich-resource languages to low-resource languages, considering that we can reuse large scale paired speech and text data in rich-resource languages to reduce cost.

LRSpeech pre-trains the TTS model θ using the paired data corpus $\mathcal{D}_{rich\_tts}$ by minimizing the loss in Eq. (6.1) and the ASR model ϕ using $\mathcal{D}_{rich\_asr}$ by minimizing the loss in Eq. (6.2).

Since the rich-resource and low-resource languages are of different phoneme or character vocabularies and speakers, LRSpeech initializes the TTS and ASR models for low-resource language with all the pre-trained parameters except the phoneme/character and speaker embeddings in the TTS model and the phoneme or character embeddings in ASR model[4] respectively. LRSpeech then fine-tunes the TTS model θ and ASR model ϕ with both D_h and D_l by minimizing the loss functions in Eqs. (6.1) and (6.2) respectively. Let θ^{ft} and ϕ^{ft} denote the TTS and ASR models after fine-tuning.

6.4.2 Dual Reconstruction

LRSpeech continues to improve the two models θ^{ft} and ϕ^{ft} using unpaired text and speech data through the process of dual learning. For each unpaired speech sequence $x \in \mathcal{X}^u$, LRSpeech converts it into a text sequence $\hat{y}$ using the latest ASR model, and then optimizes the latest TTS model to reconstruct x from $\hat{y}$. For each unpaired text sequence $y \in \mathcal{Y}^u$, LRSpeech converts it into a speech sequence $\hat{x}$ using the latest TTS model, and then optimizes the latest ASR model to reconstruct y from $\hat{x}$.

LRSpeech further introduces some specific designs in the loop of dual reconstruction to support multi-speaker TTS and ASR.

- Different from [16, 22] that only support a single speaker in both TTS and ASR models, LRSpeech supports multi-speaker speech synthesis and recognition in dual reconstruction. LRSpeech randomly chooses a speaker ID and synthesizes speech of this speaker given a text sequence, which can benefit the training of the multi-speaker ASR model. Furthermore, the ASR model converts multi-speaker speech into text, which can help the training of the multi-speaker TTS model.
- Since multiple-speaker low-quality unpaired speech data are much easier to obtain than high-quality single-speaker unpaired speech data, enabling the TTS and ASR models to utilize unseen speakers' unpaired speech in dual reconstruction can make LRSpeech more robust and scalable. Compared to ASR, it is more challenging to synthesize speech for unseen speakers. To this end, LRSpeech performs dual reconstruction in two phases, each with different kinds

[4] ASR models do not need speaker embeddings.

of unpaired data: (1) In the first phase, LRSpeech only uses the unpaired speech from speakers that appear in the paired training data. (2) In the second phase, LRSpeech adds unpaired speech from speakers that do not show up in the paired training data. As the ASR model can naturally support unseen new speakers, through dual reconstruction it enables the TTS model to synthesize speech for unseen new speakers.

Let θ^{dr} and ϕ^{dr} denote the TTS and ASR models after dual learning.

6.4.3 Knowledge Distillation

In term of quality, the TTS and ASR models θ^{dr} and ϕ^{dr} we currently get are far from ready for industrial deployment. There are several issues to address: (1) While the TTS model can support multiple speakers, the quality of the synthesized speech of the target speaker is not good enough and needs further improvement; (2) The synthesized speech by the TTS models still suffer from word skipping and repeating; (3) The accuracy of the ASR model needs to be further improved. Therefore, LRSpeech further leverages knowledge distillation [13, 27] to improve the TTS and ASR models.

The knowledge distillation process for TTS consists of three steps:

- For each unpaired text sequence $y \in \mathcal{Y}^u$, synthesize a speech sequence for the target speaker using the TTS model θ^{dr} together with the embedding of the target TTS speaker. Doing so we construct a single-speaker (i.e., the target TTS speaker) pseudo paired corpus $\mathcal{D}(\mathcal{Y}^u)$.
- Remove those pseudo text speech pairs whose synthesized speech has word skipping and repeating issues from $\mathcal{D}(\mathcal{Y}^u)$.
- Use the filtered corpus $\mathcal{D}(\mathcal{Y}^u)$ to train a new TTS model θ^{kd} dedicated to the target speaker.

In the first step, the speech in the constructed corpus $\mathcal{D}(\mathcal{Y}^u)$ contains only the target speaker for speech synthesis, which is different from the multi-speaker corpus $\mathcal{D}(\mathcal{Y}^u)$ used in Sect. 6.4.2. Since the speech generated by the TTS model θ^{dr} in the first step has word skipping and repeating issues, in the second step, LRSpeech removes the synthesized speech which has word skipping and repeating issues, and trains a new TTS model θ^{kd} on accurate text and speech pairs. In this way, the word skipping and repeating problem can be largely avoided.

Note that unpaired text and low-quality multi-speaker unpaired speech are both available for ASR.[5] LRSpeech leverages both the ASR model θ^{dr} and TTS model ϕ^{dr} to synthesize data for ASR distillation:

[5]High-quality single-speaker unpaired speech data is costly to collect while unpaired text data is easy to obtain. Therefore, LRSpeech only uses the TTS model to synthesize speech for unpaired text for TTS distillation.

Table 6.1 Statistics of training data used by LRSpeech. $\mathcal{D}_h$ denotes the high-quality paired data from the target speaker. $\mathcal{D}_l$ denotes multi-speaker low-quality paired data (50 speakers). $\mathcal{X}^u_{\text{seen}}$ denotes multi-speaker low-quality unpaired speech data (50 speakers), where speakers are seen in the paired training data $\mathcal{D}_l$. $\mathcal{X}^u_{\text{unseen}}$ denotes multi-speaker low-quality unpaired speech data (50 speakers), where speakers are unseen in the paired training data $\mathcal{D}_l$. $\mathcal{Y}^u$ denotes unpaired text data

Notation	Quality	Type	Dataset	#Samples
$\mathcal{D}_h$	High	Paired	LJSpeech [11]	50 (5 min)
$\mathcal{D}_l$	Low	Paired	LibriSpeech [18]	1000 (3.5 h)
$\mathcal{X}^u_{\text{seen}}$	Low	Unpaired	LibriSpeech	2000 (7 h)
$\mathcal{X}^u_{\text{unseen}}$	Low	Unpaired	LibriSpeech	5000 (14 h)
$\mathcal{Y}^u$	/	Unpaired	News-crawl	20,000

- For each unpaired speech $x \in \mathcal{X}^u$, generate the corresponding text using the ASR model ϕ^{dr}. Doing so we construct a pseudo corpus $\mathcal{D}(\mathcal{X}^u)$.
- For each unpaired text $y \in \mathcal{Y}^u$, synthesize the corresponding speech of multiple speakers using the TTS model θ^{dr}. Doing so we construct a pseudo corpus $\mathcal{D}(\mathcal{Y}^u)$.
- Combine $\mathcal{D}(\mathcal{X}^u)$ and $\mathcal{D}(\mathcal{Y}^u)$ as well as the single-speaker high-quality paired data $\mathcal{D}_h$ and multi-speaker low-quality paired data $\mathcal{D}_l$ to train a new ASR model ϕ^{kd}.

6.4.4 *Performance of LRSpeech*

In terms of network architectures, LRSpeech adopts Transformer based encoder-decoder framework for both TTS and ASR models. Xu et al. [34] show that dual reconstruction together with pre-training (on rich-resource languages) and post-distillation can significantly improve the TTS quality and ASR accuracy. Especially, with only 5-min high-quality speech (plus corresponding text) from the target speaker, LRSpeech learns a TTS model that meets the quality requirement of commercial speech services. The statistics of training data are shown in Table 6.1.

6.5 Dual Learning for Non-native Speech Recognition

6.5.1 *The Problem of Non-native Speech Recognition*

While latest automatic speech recognition (ASR) systems achieve very high accuracy for native speakers [33], the recognition accuracy decreases significantly when they are used by a non-native speaker of the language to be recognized. A simple and straightforward approach to improve accuracy for non-native speakers would be to train an ASR model for a specific language and nationality/ethnic group of non-native speakers of that language. However, training a high-accuracy ASR model needs large scale paired speech-text data, which is costly. Especially, since the English pronunciation of Chinese native speakers would be very different from that of Japanese native speakers, even if we only consider the recognition of English for non-native speakers, we would need to train multiples ASR models for different groups of non-native English speakers, i.e., each group is the native speakers of a certain language. Therefore, we need to train n^2 ASR models for n languages. Even if we only consider top 100 languages, this means 10,000 ASR models, and it is unaffordable for us to collect large scale labeled speech-text data for each model.

In contrast to paired datasets, unpaired datasets are both much easier to collect and larger in size for many ethnic groups speaking a second language. Therefore, Radzikowski et al. [21] propose to leverage two kinds of unpaired data, speech samples without corresponding transcripts and text corpora without corresponding speech samples, through dual learning to train ASR models for the recognition of non-native speech, and exploit the fact that speech recognition and speech synthesis are in the dual form.

The problem setting of non-native speech recognition studied in [21] is very similar to the problem of low-resource speech synthesis and speech recognition studied in [22]. Both use a small set of paired speech-text data and large scale of unpaired speech and text data. Their goals are slightly different: Radzikowski et al. [21] focus on building an ASR model while Ren et al. [22] focus on building both a TTS and an ASR models. Their methods are a bit more different. In the next sub section, we first introduce the method proposed in [21] and then compare with the method in [22].

6.5.2 The Method Based on the Dual Reconstruction Principle

The non-native speech recognition system built in [21] leverages four models:

- The first model is a text language model M_T, which is trained with a text corpus (without paired speech data). This model serves two purposes: (1) to generate a new text sentence and (2) to estimate the likelihood score of a given text sentence, i.e., how likely a given text sentence is a natural sentence.
- The second model is an acoustic model M_S, which is trained with an unlabeled non-native speech recording corpus (without paired text data). This model also serves two purposes: (1) to generate a new speech sequence (sound wave) and (2) to estimate the likelihood score of a given speech sequence, i.e., how likely a given speech sequence is a human recording.
- The third model is a speech recognition model M_{S2T}, which can recognize text from a given speech sequence.
- The fourth model is speech synthesis model M_{T2S}, which can generate a speech sequence for a given text sentence.

There are two training stages: the pre-training stage and the dual learning stage. In the pre-training stage, the text language model M_T and acoustic model M_S are independently trained using the unlabeled text corpus and the unlabeled speech corpus separately. The speech recognition model M_{S2T} and the speech synthesis model M_{T2S} can be trained in this stage using the small set of paired speech text corpus. If doing so, then the dual learning stage is warm started with the pre-trained M_{S2T} and M_{T2S}.

In the dual learning stage, the text language model M_T and acoustic model M_S are fixed and used to help the training of the speech recognition model M_{S2T} and speech synthesis model M_{T2S}. That is, only M_{S2T} and M_{T2S} are trainable and updated using unlabeled speech and text data in the second stage. The training process exactly follows the dual reconstruction principle (see Sect. 4.2) to minimize the reconstruction error of a speech sequence or text sentence through a closed loop as illustrated in Fig. 4.1. There are two kinds of loops, one started from an unlabeled text sentence, and the other one started from an unlabeled speech sequence. There are four steps in each loop. We take the loop starting with an text sequence as an example:

1. The loop begins from an unlabeled sentence t, which can (1) come from the unlabeled text corpus or (2) be generated by the text language model M_T.[6]
2. The speech synthesis model M_{T2S} takes t as input and generates a speech sequence $s = M_{T2S}(t)$.
3. The acoustic model M_S scores the speech sequence:

$$r^{im} = M_S(s).$$

   r^{im} represents how likely s is a human speech sequence.
4. The speech recognition model M_{S2T} converts s to text. Actually, we calculate the likelihood score that the original text sentence t can be reconstructed from s:

$$r^{re} = \log P(t|s; M_{S2T}).$$

Through this loop, we collect two feedback signals r^{im} and r^{re}, which are then used to update the two models M_{S2T} and M_{T2S} using policy gradient methods. The detailed algorithm can be found in [21].

While both the speech recognition model and the speech synthesis model are trained and improved after dual learning, for the purpose of non-native speech recognition, Radzikowski et al. pay most attention to the recognition model, and just treat the synthesis model as a by-product of the training process.

Both Radzikowski et al. [21] and Ren et al. [22] (Sect. 6.3) explore dual learning for low-resource speech processing. They differ from each other in a few aspects:

[6] While Radzikowski et al. [21] adopt the first option, we believe the second option will work equally well (if not better).

- In terms of tasks, Radzikowski et al. focus on speech recognition, while Ren et al. focus on both speech recognition and speech synthesis.
- For training methods, Radzikowski et al. mainly rely on dual learning; Ren et al. also leverage denoising auto-encoding (see Sect. 6.3.1) in addition to dual learning.
- For network architectures, Radzikowski et al. directly adopt RNN/LSTM, while Ren et al. enhance the Transformer model with bidirectional sequence modeling.

6.6 Beyond Speech Processing

So far we have covered natural language tasks in Chap. 4, computer vision tasks in Chap. 5, and speech related tasks in this chapter. In this section, we briefly introduce several tasks beyond natural language, computer vision, and speech processing.

In recommendation systems, the item ranking task, which aims at ranking a set of items based on users' preferences, and the user retrieval task, which aims to find a set of potential users for an item, are in the dual form. Zhang and Yang [38] propose a dual learning based ranking framework to jointly learn users' preferences over items and items' preferences over users by minimizing a pairwise ranking loss. Effective feedback signals are generated from the closed loop formed by the two dual tasks. Through a reinforcement learning process, the models for the two tasks are iteratively updated to catch relations between the item recommendation task and the user retrieval task.

User identity linking refers to the task of linking multiple IDs across social networks/platforms, i.e., detecting whether an ID in Facebook and another ID in Twitter are the same user. Zhou et al. [39] propose a dual learning based approach, which jointly learns two mapping functions: the primal function maps an ID in social network $\mathcal{X}$ to an ID in social network $\mathcal{Y}$, and the dual function maps an ID in social network $\mathcal{Y}$ to an ID in $\mathcal{X}$. They argue that for two good mapping functions, if an ID $x \in \mathcal{X}$ is linked/mapped to an ID $y \in \mathcal{Y}$ by the primal function, then y should be linked to x by the dual function. That is, the principle of dual reconstruction should hold in this problem. They show that through dual learning, not only unlabeled nodes can be exploited, but also the mapping among multiple networks can get improved through the reinforcement learning procedure.

Semi-Supervised Learning (SSL) is a popular approach to address the shortage of labeled data. A challenge for SSL is how to safely make use of the unlabeled data. Gan et al. [10] employ dual learning to estimate the safety or risk of unlabeled data samples and propose DuAL Learning-based sAfe Semi-supervised learning (DALLAS). To safely exploit unlabeled data, DALLAS first utilizes the primal model obtained by dual learning to classify each unlabeled instance and then uses the dual model to reconstruct the unlabeled instances from the output of the primal model. The risk of using an unlabeled instance is measured by (1) the reconstruction error between the original instance and the reconstructed instance and (2) the consistency of prediction results (using the primal model) of the original instance and the reconstructed unlabeled instance. If the error is small and the predictions are consistent, the unlabeled instance is likely to be safe to use. Otherwise, the instance is likely to be risky and we should be careful while using this instance. Similar idea is adopted for electroencephalogram (EEG) classification [23].

This kind of safety justification also supports the advantage of the dual reconstruction principle over back translation (see Sect. 4.1.2) in machine translation. Monolingual data in machine translation is often noisy. For a monolingual sentence y in the target language, the translation $\hat{x}$ of the dual (target-to-source) model introduces further noise due to the imperfectness of the dual model. Thus, directly adding such a pseudo pair $(\hat{x}, y)$ into the bilingual data for model training brings noise and consequently might hurt the performance of final models. In contrast, in dual learning such as the DualNMT algorithm, we use the primal model to measure the quality of $\hat{x}$: larger the reconstruction probability $P(y|\hat{x})$, better the quality of the pseudo pair $(\hat{x}, y)$. Thus, we can differentiate different unlabeled sentences in dual learning according to the dual reconstruction probability or error, and better control the risk while using noisy unlabeled data.

References

1. Arik, S., Chen, J., Peng, K., Ping, W., & Zhou, Y. (2018). Neural voice cloning with a few samples. In *Advances in Neural Information Processing Systems 31: Annual Conference on Neural Information Processing Systems 2018, NeurIPS 2018, 3–8 December 2018, Montréal* (pp. 10040–10050).
2. Bengio, S., Vinyals, O., Jaitly, N., & Shazeer, N. (2015). Scheduled sampling for sequence prediction with recurrent neural networks. In *Advances in Neural Information Processing Systems* (pp. 1171–1179).
3. Chan, W., Jaitly, N., Le, Q., & Vinyals, O. (2016). Listen, attend and spell: A neural network for large vocabulary conversational speech recognition. In *2016 IEEE International Conference on Acoustics, Speech and Signal Processing (ICASSP)* (pp. 4960–4964). Piscataway, NJ: IEEE.
4. Chen, Y.-C., Shen, C.-H., Huang, S.-F., & Lee, H.-Y. (2018). Towards unsupervised automatic speech recognition trained by unaligned speech and text only. Preprint. arXiv:1803.10952.
5. Chen, Y.-C., Shen, C.-H., Huang, S.-F., Lee, H.-Y., & Lee, L.-S. (2018). Almost-unsupervised speech recognition with close-to-zero resource based on phonetic structures learned from very small unpaired speech and text data. Preprint. arXiv:1810.12566.
6. Chen, Y., Assael, Y., Shillingford, B., Budden, D., Reed, S., Zen, H., et al. (2018). Sample efficient adaptive text-to-speech. In *International Conference on Learning Representations*.
7. Chiu, C.-C., Sainath, T. N., Wu, Y., Prabhavalkar, R., Nguyen, P., Chen, Z., et al. (2018). State-of-the-art speech recognition with sequence-to-sequence models. In *2018 IEEE International Conference on Acoustics, Speech and Signal Processing (ICASSP)* (pp. 4774–4778). Piscataway, NJ: IEEE.
8. Chuangsuwanich, E. (2016). *Multilingual Techniques for Low Resource Automatic Speech Recognition.*. Technical report, Cambridge, MA: Massachusetts Institute of Technology.
9. Dalmia, S., Sanabria, R., Metze, F., & Black, A. W. (2018). Sequence-based multi-lingual low resource speech recognition. Preprint. arXiv:1802.07420.
10. Gan, H., Li, Z., Fan, Y., & Luo, Z. (2017). Dual learning-based safe semi-supervised learning. *IEEE Access, 6*, 2615–2621.
11. Ito, K. (2017). The LJ speech dataset. https://keithito.com/LJ-Speech-Dataset/

12. Jia, Y., Zhang, Y., Weiss, R. J., Wang, Q., Shen, J., Ren, F., et al. (2018). Transfer learning from speaker verification to multispeaker text-to-speech synthesis. In *Advances in Neural Information Processing Systems 31: Annual Conference on Neural Information Processing Systems 2018, NeurIPS 2018, 3–8 December 2018, Montréal* (pp. 4485–4495).
13. Kim, Y., & Rush, A. M. (2016). Sequence-level knowledge distillation. In *Proceedings of the 2016 Conference on Empirical Methods in Natural Language Processing* (pp. 1317–1327).
14. Kuhl, P. K., Conboy, B. T., Coffey-Corina, S., Padden, D., Rivera-Gaxiola, M., & Nelson, T. (2008). Phonetic learning as a pathway to language: New data and native language magnet theory expanded (NLM-e). *Philosophical Transactions of the Royal Society B: Biological Sciences, 363*(1493), 979–1000.
15. Lample, G., Conneau, A., Denoyer, L., & Ranzato, M. A. (2018). Unsupervised machine translation using monolingual corpora only. In *Sixth International Conference on Learning Representations, ICLR 2018.*
16. Liu, A. H., Tu, T., Lee, H. Y., & Lee, L. S. (2019). Towards unsupervised speech recognition and synthesis with quantized speech representation learning. Preprint. arXiv:1910.12729.
17. Liu, D.-R., Chen, K.-Y., Lee, H.-Y., & Lee, L.-S. (2018). Completely unsupervised phoneme recognition by adversarially learning mapping relationships from audio embeddings. Preprint. arXiv:1804.00316.
18. Panayotov, V., Chen, G., Povey, D., & Khudanpur, S. (2015). Librispeech: An ASR corpus based on public domain audio books. In *2015 IEEE International Conference on Acoustics, Speech and Signal Processing (ICASSP)* (pp. 5206–5210). Piscataway, NJ: IEEE.
19. Ping, W., Peng, K., & Chen, J. (2019). Clarinet: Parallel wave generation in end-to-end text-to-speech. In *International Conference on Learning Representations.*
20. Ping, W., Peng, K., Gibiansky, A., Arik, S. O., Kannan, A., Narang, S., et al. (2018). Deep voice 3: Scaling text-to-speech with convolutional sequence learning. In *International Conference on Learning Representations.*
21. Radzikowski, K., Nowak, R., Wang, L., & Yoshie, O. (2019). Dual supervised learning for non-native speech recognition. *EURASIP Journal on Audio, Speech, and Music Processing, 2019*(1), 3, 2019.
22. Ren, Y., Tan, X., Qin, T., Zhao, S., Zhao, Z., & Liu, T. Y. (2019). Almost unsupervised text to speech and automatic speech recognition. In *International Conference on Machine Learning* (pp. 5410–5419).
23. She, Q., Zou, J., Luo, Z., Nguyen, T., Li, R., & Zhang, Y. (2020). Multi-class motor imagery EEG classification using collaborative representation-based semi-supervised extreme learning machine. *Medical & Biological Engineering & Computing, 58*(9), 2119–2130.
24. Shen, J., Pang, R., Weiss, R. J., Schuster, M., Jaitly, N., Yang, Z., et al. (2018). Natural TTS synthesis by conditioning WaveNet on mel spectrogram predictions. In *2018 IEEE International Conference on Acoustics, Speech and Signal Processing (ICASSP)* (pp. 4779–4783). Piscataway, NJ: IEEE.
25. Shen, S., Cheng, Y., He, Z., He, W., Wu, H., Sun, M., et al. (2016). Minimum risk training for neural machine translation. In *Proceedings of the 54th Annual Meeting of the Association for Computational Linguistics, ACL 2016, August 7–12, 2016, Berlin, Volume 1: Long Papers.*
26. Sun, H., Tan, X., Gan, J.-W., Liu, H., Zhao, S., Qin, T., et al. (2019). Token-level ensemble distillation for grapheme-to-phoneme conversion. *Proceedings of the Interspeech 2019* (pp. 2115–2119).
27. Tan, X., Ren, Y., He, D., Qin, T., Zhao, Z., & Liu, T. Y. (2019). Multilingual neural machine translation with knowledge distillation. In *International Conference on Learning Representations.*
28. Tjandra, A., Sakti, S., & Nakamura, S. (2017). Listening while speaking: Speech chain by deep learning. In *Automatic Speech Recognition and Understanding Workshop (ASRU), 2017 IEEE* (pp. 301–308). Piscataway, NJ: IEEE.
29. Tjandra, A., Sakti, S., & Nakamura, S. (2018). Machine speech chain with one-shot speaker adaptation. *Proceedings of the Interspeech 2018* (pp. 887–891).

30. Wang, Y., Skerry-Ryan, R. J., Stanton, D., Wu, Y., Weiss, R. J., Jaitly, N., et al. (2017). Tacotron: Towards end-to-end speech synthesis. Preprint. arXiv:1703.10135.
31. Wang, Y., Stanton, D., Zhang, Y., Skerry-Ryan, R. J., Battenberg, E., Shor, J., et al. (2018). Style tokens: Unsupervised style modeling, control and transfer in end-to-end speech synthesis. In *Proceedings of the 35th International Conference on Machine Learning, ICML 2018, Stockholmsmässan, Stockholm, Sweden, July 10–15, 2018* (pp. 5167–5176).
32. Wind, J. (1989). The evolutionary history of the human speech organs. *Studies in Language Origins, 1*, 173–197.
33. Xiong, W., Droppo, J., Huang, X., Seide, F., Seltzer, M., Stolcke, A., et al. (2016). Achieving human parity in conversational speech recognition. Preprint. arXiv:1610.05256.
34. Xu, J., Tan, X., Ren, Y., Qin, T., Li, J., Zhao, S., et al. (2020). LRSpeech: Extremely low-resource speech synthesis and recognition. In *Proceedings of the 26th ACM SIGKDD International Conference on Knowledge Discovery and Data Mining*.
35. Yang, X., Li, J., & Zhou, X. (2018). A novel pyramidal-FSMN architecture with lattice-free MMI for speech recognition. Preprint. arXiv:1810.11352.
36. Yeh, C.-K., Chen, J., Yu, C., & Yu, D. (2019). Unsupervised speech recognition via segmental empirical output distribution matching. In *ICLR*.
37. Zhang, S., Lei, M., Yan, Z., & Dai, L. (2018). Deep-FSMN for large vocabulary continuous speech recognition. Preprint. arXiv:1803.05030.
38. Zhang, Z., & Yang, J. (2018). Dual learning based multi-objective pairwise ranking. In *2018 International Joint Conference on Neural Networks (IJCNN)* (pp. 1–7). Piscataway, NJ: IEEE.
39. Zhou, F., Liu, L., Zhang, K., Trajcevski, G., Wu, J., & Zhong, T. (2018). Deeplink: A deep learning approach for user identity linkage. In *IEEE INFOCOM 2018-IEEE Conference on Computer Communications* (pp. 1313–1321). Piscataway, NJ: IEEE.
40. Zhou, S., Xu, S., & Xu, B. (2018). Multilingual end-to-end speech recognition with a single transformer on low-resource languages. Preprint. arXiv:1806.05059.

Part III
The Probabilistic Principle

Structure duality can also be interpreted from the perspective of probability. In this part, we introduce several dual learning algorithms that exploit structure duality based on different probability equations and for different settings:

- dual supervised learning that leverages the joint probability constraint to enhance learning from labeled data;
- dual inference that leverages the conditional probability constraint in inference;
- dual semi-supervised learning that leverages the marginal probability constraint to learn from unlabeled data.

Chapter 7
Dual Supervised Learning

7.1 The Joint-Probability Principle

In previous chapters, we show how structure duality can be leveraged to enable learning from unlabeled data based on the principle of dual reconstruction. Structure duality can imply a lot more than the principle of dual reconstruction. Here we consider the setting of supervised learning, i.e., how structure duality can enhance the learning from the labeled data.

We first define some notations. The primal task takes a sample from space $\mathcal{X}$ as input and maps to space $\mathcal{Y}$, and the dual task takes a sample from space $\mathcal{Y}$ as input and maps to space $\mathcal{X}$. Using the language of probability, the primal task learns a conditional distribution $P(y|x;\theta_{XY})$ parameterized by θ_{XY}, and the dual task learns a conditional distribution $P(x|y;\theta_{YX})$ parameterized by θ_{YX}, where $x \in \mathcal{X}$ and $y \in \mathcal{Y}$.

For any $x \in \mathcal{X}$, $y \in \mathcal{Y}$, the joint probability $P(x, y)$ can be computed in two equivalent ways: $P(x, y) = P(x)P(y|x) = P(y)P(x|y)$. If the two models θ_{XY} and θ_{YX} are perfect, their parameterized conditional distributions should satisfy the following equality:

$$P(x)P(y|x;\theta_{XY}) = P(y)P(x|y;\theta_{YX}).\forall x \in \mathcal{X}, y \in \mathcal{Y}. \tag{7.1}$$

This equation defines a relationship between the primal model θ_{XY} and the dual model θ_{YX} from the perspective of probability, and we call it the *joint-probability principle*.

However, if the two models (conditional distributions) are trained separately by minimizing their own loss functions (as in the major practice of machine learning), there is no guarantee that the above equation will hold. The basic idea of dual supervised learning is to jointly train the two models θ_{XY} and θ_{YX} by minimizing their loss functions subject to the constraint of Eq. (7.1). By doing so, the intrinsic probabilistic connection between θ_{YX} and θ_{XY} are explicitly strengthened, which is supposed to push the learning process towards the right direction.

T. Qin, *Dual Learning*, https://doi.org/10.1007/978-981-15-8884-6_7

7.2 The Algorithm of Dual Supervised Learning

In this section, we formulate the problem of dual supervised learning (DSL), describe an algorithm [33] for DSL, and discuss its connections with existing learning schemes and its application scope.

Let $\mathcal{D}$ denote a set of training pairs (x, y) with $x \in \mathcal{X}$ and $y \in \mathcal{Y}$. Let θ_{XY} denote the parameters of the primal model mapping from $\mathcal{X}$ to $\mathcal{Y}$, and θ_{YX} the parameters of the dual model mapping from $\mathcal{Y}$ to $\mathcal{X}$. In conventional supervised learning, the two models are trained by minimizing the empirical risk over training data, e.g., negative log likelihood in deep learning:

$$\min_{\theta_{XY}} -\frac{1}{|\mathcal{D}|} \sum_{(x,y)\in\mathcal{D}} \log P(y|x;\theta_{XY});$$

$$\min_{\theta_{YX}} -\frac{1}{|\mathcal{D}|} \sum_{(x,y)\in\mathcal{D}} \log P(x|y;\theta_{YX}).$$

Correspondingly, we induce the following prediction functions for the primal and dual tasks:

$$f(x;\theta_{XY}) \triangleq \arg\max_{y'\in\mathcal{Y}} P(y'|x;\theta_{XY}),$$

$$g(y;\theta_{YX}) \triangleq \arg\max_{x'\in\mathcal{X}} P(x'|y;\theta_{YX}).$$

Clearly, the parameters θ_{XY} and θ_{YX} of two perfect models should satisfy the constraint described by the joint-probability principle in Eq. (7.1). Unfortunately, in conventional supervised learning, the primal and the dual models are trained independently and separately, and the constraint of joint probability is not considered in training. Thus, there is no guarantee that the learned models can satisfy the constraint.

To tackle this problem, dual supervised learning [33] jointly trains the two models and explicitly reinforce the constraint of joint probability for all training pairs (x, y), which results in the following multi-objective optimization:

$$\begin{aligned}
&\text{objective 1: } \min_{\theta_{XY}} \frac{1}{|\mathcal{D}|} \sum_{(x,y)\in\mathcal{D}} \ell_1(f(x;\theta_{XY}), y),\\
&\text{objective 2: } \min_{\theta_{YX}} \frac{1}{|\mathcal{D}|} \sum_{(x,y)\in\mathcal{D}} \ell_2(g(y;\theta_{YX}), x),\\
&\text{s.t. } P(x)P(y|x;\theta_{XY}) = P(y)P(x|y;\theta_{YX}), \forall(x,y)\in\mathcal{D},
\end{aligned} \tag{7.2}$$

where $P(x)$ and $P(y)$ are the marginal distributions, and $\ell_1()$ and $\ell_2()$ denote the loss functions for the primal and dual tasks respectively. In practical applications, the ground-truth marginal distributions $P(x)$ and $P(y)$ are usually not available. As an alternative, empirical marginal distributions $\hat{P}(x)$ and $\hat{P}(y)$ can be employed to fulfill the constraint in Eq. (7.2).

To solve the above optimization problem, following the common practice in constrained optimization, Xia et al. [33] introduce Lagrange multipliers and convert the equality constraint of joint probability to the third objective, in addition the two in Eq. (7.2). First, one converts the joint probability constraint into the following regularization term:

$$\ell_{\text{dsl}} = (\log \hat{P}(x) + \log P(y|x; \theta_{XY}) - \log \hat{P}(y) - \log P(x|y; \theta_{YX}))^2. \quad (7.3)$$

Then, one trains the models of the two tasks by minimizing the weighted combination between the original loss functions and the above regularization term. The algorithm is shown in Algorithm 1.

Algorithm 1 Dual supervise learning algorithm

Require: : Marginal distributions $\hat{P}(x)$ and $\hat{P}(y)$; Lagrange parameters λ_{XY} and λ_{YX}; optimizers Opt_1 and Opt_2;

repeat

Sample a minibatch of m pairs $\{(x_j, y_j)\}_{j=1}^m$;

Calculate the gradients as follows:

$$G_f = \nabla_{\theta_{XY}}(1/m)\sum_{j=1}^m\big[\ell_1(f(x_j;\theta_{XY}), y_j) + \lambda_{XY}\ell_{\text{dsl}}(x_j, y_j; \theta_{XY}, \theta_{YX})\big];$$

$$G_g = \nabla_{\theta_{YX}}(1/m)\sum_{j=1}^m\big[\ell_2(g(y_j;\theta_{YX}), x_j) + \lambda_{YX}\ell_{\text{dsl}}(x_j, y_j; \theta_{XY}, \theta_{YX})\big];$$

Update the parameters of f and g:

$$\theta_{XY} \leftarrow Opt_1(\theta_{XY}, G_f), \theta_{YX} \leftarrow Opt_2(\theta_{YX}, G_g).$$

until models converge

In the algorithm, the choice of optimizers Opt_1 and Opt_2 is quite flexible. One can choose different optimizers such as Adadelta [37], Adam [16], or SGD for different tasks, depending on common practice in the specific task and personal experiences.

While ℓ_{dsl} can be regarded as a regularization term, it is data dependent, which makes DSL different from Lasso [30] or SVM [6], where the regularization term is data independent. More accurately speaking, in DSL, the regularization term depends on both the model parameters (θ_{XY} and θ_{YX}) and the training pairs $(x, y) \in \mathcal{D}$, while the regularization term in Lasso or SVM only depends on model parameters. Furthermore, since DSL conducts joint training, the primal and dual models regularize each other during training.

We would like to point out that there are several requirements to apply DSL to a certain scenario:

- Structure duality should exist for the two tasks.
- Both the primal and dual models should be trainable.
- $\hat{P}(X)$ and $\hat{P}(Y)$ in Eq. (7.3) should be available.

If these conditions are not satisfied, DSL might not work very well. Fortunately, as we will discuss in the next section, many machine learning tasks related to image, speech, and text satisfy these conditions.

7.3 Applications

Dual supervised learning has been studied in many applications. In this section, we selectively introduce several representative ones. Roughly speaking, to apply the above DSL algorithm, one need to specify the empirical marginal distributions $\hat{P}(X)$ and $\hat{P}(Y)$, the function class (e.g., the structure for neural networks) of the primal and dual models, and the detailed training procedure.

7.3.1 Neural Machine Translation

Xia et al. [33] apply DSL to neural machine translation and study whether it can improve the translation quality based on the joint-probability principle. They perform experiments on three pairs of dual tasks:[1] English↔French (En↔Fr), English↔Germany (En↔De), and English↔Chinese (En↔Zh).

Marginal Distributions *$\hat{P}(x)$ and $\hat{P}(y)$* The LSTM-based language modeling approach [21, 29] is used to characterize the empirical marginal distribution of a sentence x, defined as $\prod_{i=1}^{T_x} P(x_i|x_{<i})$, where x_i is the i-th word/token in x, T_x denotes the number of words/tokens in x, and the index $< i$ is the set $\{1, 2, \cdots, i-1\}$.

Model GRU based encoder-decoder framework is used to implement the translation models, which is the same as [3, 15]. The vocabulary sizes of the source and target languages are set to 30k, 50k, and 30k for En↔Fr, En↔De, and En↔Zh, respectively. The out-of-vocabulary words are replaced by a special token UNK. Of course, as neural machine translation has made great success in rent years, the model settings used in [33] can definitely be improved. For example, one can use Transformer [31] to replace the GRU based model, and use sub word units like BPE [26] to better handle out-of-vocabulary words.

---

[1] Note that the two translation tasks in a language pair are symmetric. They play the same role in dual supervised learning. Consequently, either one of the two tasks can be viewed as the primal task while the other as the dual task.

Training Procedure The two models in DSL (θ_{XY} and θ_{YX}) are initialized by using two pre-trained models, which is generated following the same process as [15]. Then, SGD is used for model training. The value of both λ_{XY} and λ_{YX} in Algorithm 1 are set as 0.01 according to empirical performance on the validation set. Note that, during the optimization process, the LSTM-based language models are fixed.

Results Experimental result [33] show significant accuracy improvements by applying DSL to machine translation: $+2.07/0.86$ points measured by BLEU scores for English↔French translation, $+1.37/0.12$ points for English↔Germen translation and $+0.74/1.69$ points on English↔Chinese.

7.3.2 Images Classification and Generation

In the domain of image processing, image classification (mapping from an image to a category label) and conditional image generation[2] (generating an image for a given category label) are in the dual form. Xia et al. [33] apply the dual supervised learning framework to these two tasks, in which image classification is treated as the primal task and conditional image generation is treated as the dual task. They conduct experiments on a public dataset CIFAR-10 [17] with 10 classes of images.

Marginal Distributions The simple uniform distribution is adopted for the marginal distribution $\hat{P}(y)$ of 10-class labels, i.e., the marginal probability of each class equals 0.1. The marginal image distribution $\hat{P}(x)$ is defined as $\prod_{i=1}^{m} P\{x_i|x_{<i}\}$, where all pixels of the image are serialized and x_i is the value of the i-th pixel of an m-pixel image. Note that the model can predict x_i only based on the previous pixels x_j with index $j < i$. PixelCNN++ is adopted to model the marginal image distribution.

Models A popular method, ResNet, is adopted for the task of image classification. In particular, a 32-layer ResNet (denoted as ResNet-32) and a 110-layer ResNet (denoted as ResNet-110) as two baselines, respectively, in order to examine the power of DSL on both relatively simple and complex models. For the task of image generation, PixelCNN++ is adopted, again. Compared with the PixelCNN++ used for modeling the marginal distribution, the difference lies in the training process: When used for image generation conditioned on a category label, PixelCNN++ takes a category label as an additional input, i.e., it tries to characterize $\prod_{i=1}^{m} \mathbb{P}\{x_i|x_{<i}, y\}$, where y is the 1-hot category vector.

[2]Note that the task of conditional image generation is different from the task of conditional image to image translation introduced in Sect. 5.4. The former one takes a category label as input and generates an image, while the latter one takes two images (one main input image and one conditional input image) from two domains as input and generates an image.

Training Procedure The primal and dual models are first initialized with the ResNet model and PixelCNN++ model pre-trained independently and separately. A 32-layer ResNet with error rate of 7.65 and a 110-layer ResNet with error rate of 6.54 are obtained as the pre-trained models for image classification. The error rates of these two pre-trained models are comparable to previously reported results [14]. A pre-trained conditional image generation model is obtained with the test *bpd*[3] of 2.94, which is the same as reported in [25]. The hyper parameters are set as $\lambda_{XY} = (30/3072)^2$ and $\lambda_{YX} = (1.2/3072)^2$.

Results As shown in [33], for image classification, DSL significantly improves classification accuracy, reducing the error rate of ResNet-32 from 7.51% to 6.82% and ResNet-110 from 6.43% to 5.40%; for conditional image generation, based on ResNet-110, DSL reduces the test bpd from 2.94 to 2.93, which is a new record on CIFAR-10 at that time.

7.3.3 Sentiment Analysis

Xia et al. [33] apply the dual supervised learning algorithm to the domain of sentiment analysis. In this domain, the primal task, sentiment classification [9, 19], is to predict the sentiment polarity label of a given sentence/paragraph; and the dual task, though not quite apparent but really existed, is sentence generation based on a sentiment polarity. Let $\mathcal{X}$ denote the set of sentences and $\mathcal{Y}$ denote the set of sentiment labels, e.g., positive and negative.

Marginal Distributions The simple uniform distribution is adopted for the marginal distribution $\hat{P}(y)$ of polarity labels, i.e., the marginal probability of positive or negative class is 0.5. The LSTM-based language models are used to model the marginal distribution $\hat{P}(x)$ of a sentence x.

Model Implementation A two-layer LSTM [9] is used as sentiment classification model. Sentence generation is based on another LSTM model, and a sentence is generated word by word sequentially.

Training Procedure AdaDelta is used to train the two baseline LSTM models, one for sentiment classification and the other for sentence generation. Then the well-trained baseline models are used to initialize the two models for DSL training with

[3] *Bits per dimension* (briefly, *bpd*) [25] is used to assess the performance of image generation. In particular, for an image x with label y, the bpd is defined as:

$$-\left(\sum_{i=1}^{N_x} \log P(x_i|x_{<i}, y)\right)/\left(N_x \log(2)\right), \tag{7.4}$$

where N_x is the number of pixels in image x. For images in the dataset CIFAR-10, N_x is 3072.

plain SGD as the optimizer. For each (x, y) pair, the hyper parameters are set as $\lambda_{xy} = (5/l_x)^2$ and $\lambda_{xy} = (0.5/l_x)^2$, where l_x is the length of x.

Results The experiments are performed on the IMDB movie review dataset,[4] which consists of 25k training and 25k test sentences. Each sentence in this dataset is associated with either a positive or a negative sentiment label. A subset of 3750 sentences is randomly sampled from the training data as the validation set for hyper parameter tuning and the remaining training data is used for model training. The results show that DSL reduces the error rate of sentence classification from 10.10% to 9.20%, and reduces the test perplexity of sentence generation from 59.19 to 58.78.

7.3.4 Question Answering and Generation

Question answering (QA) and question generation (QG) are two fundamental tasks in natural language processing [20] and they play an important role in search engines and conversational bots. Sun et al. [28] apply dual supervised learning for QA and QG.

There are different kinds of QA tasks in natural language processing community. An important format of QA, answer sentence selection [34], is considered in [28], which takes a question q and a list of candidate answer sentences $A = \{a_1, a_2, \ldots, a_{|A|}\}$ as input, and outputs one answer sentence a_i from the candidate list which has the largest probability to be the answer. This QA task is a typical ranking problem. Let $f(a, q; \theta_{QA})$ denote a QA model parameterized by θ_{QA} and its output is a real-valued matching score between a question q and a candidate answer a.

Given a sentence a as input, the QG task aims to generate a question q which could be answered by a. Since both a and q are sentences, QG is a standard sequence-to-sequence learning problem. Let $P(q|a; \theta_{AQ})$ denote a QG model parameterized by θ_{AQ} and its output is the probability of generating a natural language question q for the input answer a.

As the QA task is actually a ranking problem, to take ranking into consideration, $P(a|q; \theta_{QA})$ is defined in a contrastive learning manner using the QA model $f(; \theta_{QA})$ together with a set of randomly sampled candidate answers (serving as negative answers for the question q):

$$P(a|q; \theta_{QA}) = \frac{\exp(f(a, q; \theta_{QA}))}{\exp(f(a, q; \theta_{QA})) + \sum_{a' \in A'} \exp(f(a', q; \theta_{QA}))}. \tag{7.5}$$

[4]The dataset can be downloaded at http://ai.stanford.edu/~amaas/data/sentiment/aclImdb_v1.tar.gz.

Then the QA specific objective is the negative log likelihood

$$\ell_1(q, a) = -\log P(a|q; \theta_{QA}), \tag{7.6}$$

where a is the correct answer of q.

Similarly, the QG specific objective is

$$\ell_2(q, a) = -\log P(q|a; \theta_{AQ}) \tag{7.7}$$

where a is the correct answer of q. Since a standard sequence to sequence model will output a probability for a pair of input and output sequences, we omit the details of the QG model here to avoid redundancy. Interested readers can find details at [28].

The third objective is the regularization term that is derived from the joint-probability principle (see Eq. (7.1)). Specifically, given a correct (q, a) pair, one needs to minimize the following loss function,

$$\begin{aligned} \ell_{\text{dsl}}(q, a; \theta_{QA}, \theta_{AQ}) = &\big[\log P(a) + \log P(q|a; \theta_{AQ}) \\ &- \log P(q) - \log P(a|q; \theta_{QA})\big]^2 \end{aligned} \tag{7.8}$$

where $P(a)$ and $P(q)$ are marginal distributions, which could be obtained through language model similar to the case of sentiment analysis in Chap. 7.3.3.

Then one can adopt Algorithm 1 to jointly train the primal and dual models through minimizing the three objectives in Eqs. (7.6)–(7.8).

Experiments on three datasets including MARCO [4], SQUAD [24], and WikiQA [34] show that DSL can significantly improve both the question answering and question generation models.

In open-domain dialogue systems, generative approaches suffer from safe responses and unnatural responses. To address the two problems, Cui et al.[8] propose the Dual Adversarial Learning (DAL) framework to generate high-quality responses. DAL leverages the duality between the query generation task and the response generation task and conducts joint training for the two tasks in the supervised setting to avoid safe responses and increase the diversity of the generated responses. In addition, it uses adversarial learning to mimic human judges and guides the system to generate natural responses.

He et al. [13] consider the probabilistic connections between the models of the visual question answering task and the visual question generation task and jointly train the two models with the joint probability constraint.

7.3.5 *Code Summarization and Generation*

Code summarization aims at generating a human readable summary in natural language that describes the functionality of a program or a piece of code. High-quality summaries are useful for applications such as code retrieval and code documentation, but are costly to manually annotate, especially for existing code. Automatic code summarization is in great demand and multiple deep neural networks based models/algorithms have been designed [1, 2, 10].

Code generation, the reverse task of code summarization, aims to automatically generate source code in some programming language like Python for natural language descriptions. It will not only reduce the workload of programmers and consequently improving their productivity, but also enable non-professional programmers to implement their ideas and unblock their creativity. Code generation has been studied in both natural language processing community and machine learning community [12, 18, 36].

Although code summarization and code generation are naturally in the dual form, most previous works study them separately. They are jointly studied only very recently under the framework of dual learning [32, 35].

7.3.5.1 Augmentation with Attention Duality

Wei et al. [32] formulate code generation and code summarization as two sequence to sequence learning problems and follow the basic idea of dual supervised learning in Eq. (7.2). That is, in addition to maximizing the likelihood of code-summary pairs in the training set, they add the duality constraint of joint probability equation as shown in Eq. (7.1). Furthermore, they observe that attention of the primal and dual models should also be symmetric and propose to enhance dual supervised learning with attention duality regularization.

Intuitively, in the primal task of code generation, if a token x_i in a code sequence x attends to a token y_j in the corresponding text summary y, then in the dual task of code summarization, the text token y_j in y should attend to the code token x_i in x. In other words, the attention in the primal and dual models should be symmetric, which is called attention duality in [32].

Consider a source code sequence x and its text summary y. Suppose there are m tokens in x and n tokens in y. Let A denote the attention matrix extracted from the primal code generation model, in which $A_{i,j}$ is the attention weight from token x_i to token y_j. Let A' denote the attention matrix extracted from the dual code summarization model, in which $A'_{i,j}$ is the attention weight from token y_i to token x_j. Let b_i denote the attention distribution in the primal code generation model that a code token x_i attends to all the text tokens in y. We have

$$b_i = softmax(A_{i,:}),$$

where $A_{i,:}$ is the i-th row of the attention matrix A. Similarly, we define

$$b'_i = softmax(A'_{:,i}),$$

where $A'_{:,i}$ is the i-th column of the attention matrix A'.

If the two models are perfect, we have $b_i = b'_i$. Obviously, without explicit constraints, it is difficult to guarantee $b_i = b'_i$. Therefore, Wei et al. introduce the Jensen–Shannon divergence [11], a symmetric measurement of similarity between two probability distributions, to penalize the disagreement between the two attention matrices:

$$l_{att}^{cg}(x, y) = \frac{1}{2m} \sum_{i=1}^{m} \left[D_{KL}(b_i \| \frac{b_i + b_i^{'}}{2}) + D_{KL}(b_i^{'} \| \frac{b_i + b_i^{'}}{2}) \right] \tag{7.9}$$

where D_{KL} is the Kullback–Leibler divergence, defined as

$$D_{KL}(p\|q) = \sum_x p(x) log \frac{p(x)}{q(x)},$$

measuring how one probability distribution p diverges from the other probability distribution q.

The attention loss l_{att}^{cg} defined above stands on the view point of source code. We can similarly define the attention loss l_{att}^{cs} from the perspective of text summaries:

$$l_{att}^{cs}(x, y) = \frac{1}{2n} \sum_{i=1}^{n} \left[D_{KL}(c_i \| \frac{c_i + c_i^{'}}{2}) + D_{KL}(c_i^{'} \| \frac{c_i + c_i^{'}}{2}) \right], \tag{7.10}$$

where

$$c_i = softmax(A'_{i,:}),$$

and

$$c'_i = softmax(A_{:,i}).$$

The total loss for a code-summary pair (x, y) becomes

$$\begin{aligned} l(x, y) = & - \log P(y|x; \theta_{XY}) - \log P(x|y; \theta_{YX}) \\ & - \lambda_1 \left(\log \hat{P}(x) + \log P(y|x; \theta_{XY}) - \log \hat{P}(y) - \log P(x|y; \theta_{YX}) \right)^2 \\ & + \lambda_2 \left(l_{att}^{cg}(x, y) + l_{att}^{cs}(x, y) \right), \end{aligned}$$

where the first line is the conventional supervised loss, the second line is the regularization term derived from the probability duality, the third line is the attention loss derived from the attention duality, and λ_1 and λ_2 are hyper-parameters.

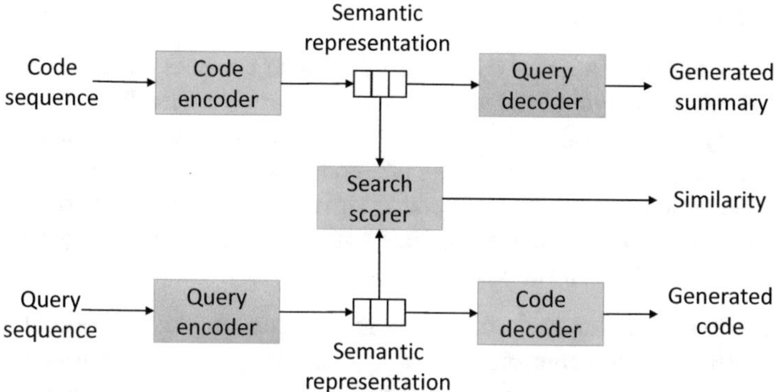

Fig. 7.1 Key components of CO3 model

The experiments in [32] show that both the joint-probability principle and attention duality lead to accuracy improvement for code generation and code summarization, and their combination leads to further improvement.

7.3.5.2 Code Retrieval, Summarization, and Generation

Code retrieval aims to find relevant source code from a set of candidates for a given natural language query. Ye et al. [35] focus on code retrieval and summarization and design an end-to-end model for both code retrieval and code summarization by introducing the code generation task and exploiting the intrinsic connection between these tasks via dual learning and multi-task learning. The designed model is called CO3 as 3 code related tasks are involved.

Figure 7.1 illustrates the high-level architecture of the CO3 model. As can be seen, there are several key components in CO3: a code encoder, a code decoder, a query encoder, a query decoder, and a similarity scorer.

- The code encoder takes a code sequence as input and outputs a set of semantic representations for the code sequence.
- The code decoder takes a set of semantic representations as inputs and outputs a code sequence.
- The query encoder takes a query/text sequence as input and outputs a set of semantic representations for the query/text sequence.
- The query decoder takes a set of semantic representations as inputs and outputs a query/text sequence.
- The similarity scorer takes two sets of semantic representations as inputs and outputs a similarity score for the two inputs.

In CO3, the code encoder and decoder share parameters, and the query encoder and decoder share parameters.

CO3 serves 3 tasks as follows:

- The code encoder followed by the query decoder converts a code sequence to a code summary, playing the role of a code summarization model.
- The query encoder followed by the code decoder converts a query/text sequence to a code sequence, playing the role of a code generation model.
- The code encoder and query encoder together with the similarity scorer can compute the similarity between a code sequence and a query sequence, playing the role of a code retrieval model.

Similar to the dual supervised learning in previous sub section, Ye et al. take the duality between code summarization and code generation into consideration and leverage the joint-probability principle between the code summarization model and the code generation model to regularize the training of CO3.

Let $\mathcal{D}$ denote the training set of code-summary pairs, θ_{cs} the parameters of the code summarization model, θ_{cg} the parameters of the code generation model, and θ_{cr} the parameters of the code retrieval model. Note that there are overlapped parameters among θ_{cs}, θ_{cg} and θ_{cr} because of parameter sharing.

CO3 adopts multitask learning and considers multiple objectives including (1) the data likelihood of code summarization, (2) the data likelihood of code generation, (3) the regularization term induced from the joint-probability principle, and (4) the objective for code retrieval. The first three objectives have already been discussed in previous sections. Here we introduce the last one for code retrieval.

Let $f(x, y; \theta_{cr})$ denote the model for code retrieval with parameters θ_{cr} which outputs a similarity score for a code-query pair (x, y).Ye et al. use the margin based ranking loss to define the training objective for code retrieval. Consider a ground-truth code-query pair (x, y) and another pair (x^{rn}, y^{rn}) with a randomly sampled code sequence x^{rn} and a randomly sampled query y^{rn}. Clearly, we should have

$$f(x, y; \theta_{cr}) > f(x^{rn}, y^{rn}; \theta_{cr}).$$

Based on this intuition, Ye et al. define the loss as

$$l_{cr}(x, y, x^{rn}, y^{rn}; \theta_{cr}) = \max\left(f(x^{rn}, y^{rn}; \theta_{cr}) + m_{cr} - f(x, y; \theta_{cr}), 0\right),$$

which means that the similarity score of a correct code-query pair should be higher than that of a random pair with a margin of at least m_{cr}.

Now we summarize the overall training objectives of CO3 as follows.

$$\min -\frac{1}{|\mathcal{D}|} \sum_{(x,y)\in\mathcal{D}} \log P(y|x; \theta_{cs})$$

$$\min -\frac{1}{|\mathcal{D}|} \sum_{(x,y)\in\mathcal{D}} \log P(x|y; \theta_{cg})$$

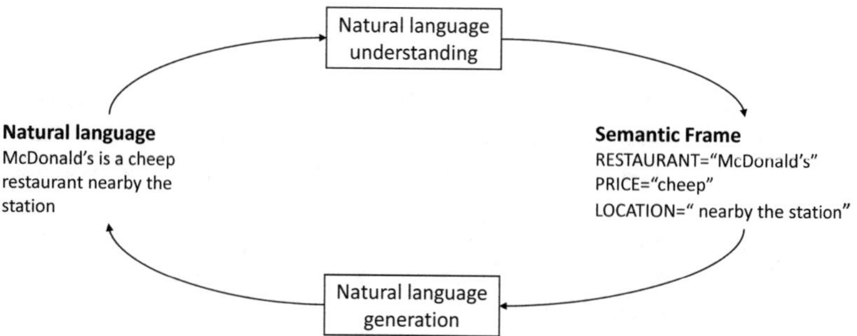

Fig. 7.2 Dual supervised learning for natural language understanding (from utterances to semantics) and natural language generation (from semantics to natural sentences)

$$\min -\frac{1}{|\mathcal{D}|} \sum_{(x,y)\in\mathcal{D}} l_{cr}(x, y, x^{rn}, y^{rn}; \theta_{cr})$$

$$\text{s.t.} \quad P(x)P(y|x, \theta_{cs}) = P(y)(P(x|y, \theta_{cg})$$

Experiments on SQL and Python datasets show that CO3 significantly improve code retrieval over state-of-the-art models, without affecting the performance in code summarization.

7.3.6 *Natural Language Understanding and Generation*

Natural language understanding (NLU), which is to extract the core semantic meaning from given utterances, and natural language generation (NLG), which is to construct corresponding sentences based on given semantics, are two fundamental tasks in building task-oriented dialogue systems. As shown in Fig. 7.2, these two tasks are in the dual form. Unfortunately, such dual relationship has not been well investigated in the supervised setting in literature. Su et al. [27] exploit the duality on top of dual supervised learning.[5]

Su et al. use a language model to model the empirical marginal distribution of natural sentences and a masked autoencoder to model the empirical marginal distribution of discrete semantic frames. Experiments on the E2E NLG challenge dataset [23], which is a crowd-sourced dataset of 50k instances in the restaurant domain, demonstrate the effectiveness of dual supervised learning for natural language understanding and generation.

[5]There is a contemporary work [7] studying dual learning for natural language understanding and generation in the semi-supervised setting, as introduced in Chap. 4.6.1.

7.4 Theoretical Analysis

Given that empirical success of dual supervised learning has been observed in many applications, it is natural to ask whether there are some theoretical evidences to support its success. Here we provide some theoretical analysis to dual supervised learning [33].

The final goal of the dual learning is to make correct predictions for the unseen test data. That is, we want to minimize the expected risk of the primal and dual models, which is defined as follows:[6]

$$R(f, g) = \mathbb{E}\left[\frac{\ell_1(f(x), y) + \ell_2(g(y), x)}{2}\right], \forall f \in \mathcal{F}, g \in \mathcal{G},$$

where $\mathcal{F} = \{f(x; \theta_{XY}); \theta_{XY} \in \Theta_{XY}\}$, $\mathcal{G} = \{g(x; \theta_{YX}); \theta_{YX} \in \Theta_{YX}\}$, Θ_{XY} and Θ_{YX} are parameter spaces, and the $\mathbb{E}$ is taken over the underlying distribution P. Besides, let $\mathcal{H}_{\text{dsl}}$ denote the product space of the two models satisfying the joint-probability constraint in Eq. (7.1).

Define the empirical risk on the n training pairs as follows.

$$R_n(f, g) = \frac{1}{n}\sum_{i=1}^{n} \frac{\ell_1(f(x_i), y_i) + \ell_2(g(y_i), x_i)}{2}$$

Following [5], we introduce Rademacher complexity for dual supervised learning, a measure for the complexity of the hypothesis.

Definition 7.1 Define the Rademacher complexity of DSL, $\mathbb{R}_n^{\text{dsl}}$, as

$$\mathbb{R}_n^{\text{dsl}} = \mathop{\mathbb{E}}_{z,\sigma}\Big[\sup_{(f,g)\in\mathcal{H}_{\text{dsl}}} \Big|\frac{1}{n}\sum_{i=1}^{n} \sigma_i\big(\ell_1(f(x_i), y_i) + \ell_2(g(y_i), x_i)\big)\Big|\Big],$$

where $z = \{z_1, z_2, \cdots, z_n\} \sim P^n$, $z_i = (x_i, y_i)$ in which $x_i \in \mathcal{X}$ and $y_i \in \mathcal{Y}$, $\sigma = \{\sigma_1, \cdots, \sigma_m\}$ are i.i.d sampled with $P(\sigma_i = 1) = P(\sigma_i = -1) = 0.5$.

Based on $\mathbb{R}_n^{\text{dsl}}$, we have the following theorem for dual supervised learning.

Theorem 7.2 ([22]) *Let $\frac{1}{2}\ell_1(f(x), y) + \frac{1}{2}\ell_2(g(y), x)$ be a mapping from $\mathcal{X} \times \mathcal{Y}$ to $[0, 1]$. Then, for any $\delta \in (0, 1)$, with probability at least $1 - \delta$, the following inequality holds for any $(f, g) \in \mathcal{H}_{dsl}$,*

$$R(f, g) - R_n(f, g) \leq 2\mathbb{R}_n^{dsl} + \sqrt{\frac{1}{2n}\ln(\frac{1}{\delta})}.$$

[6] For simplicity, the parameters θ_{XY} and θ_{YX} in the models will be omitted when the context is clear.

Similarly, we define the Rademacher complexity for the standard supervised learning $\mathbb{R}_n^{\mathrm{sl}}$ under our framework by replacing the $\mathcal{H}_{\mathrm{dsl}}$ in Definition 7.1 by $\mathcal{F} \times \mathcal{G}$. With probability at least $1 - \delta$, the generation error bound of supervised learning is smaller than $2\mathbb{R}_n^{\mathrm{sl}} + \sqrt{\frac{1}{2n} \ln(\frac{1}{\delta})}$.

Since $\mathcal{H}_{\mathrm{dsl}} \in \mathcal{F} \times \mathcal{G}$, by the definition of Rademacher complexity, we have $\mathbb{R}_n^{\mathrm{dsl}} \leq \mathbb{R}_n^{\mathrm{sl}}$. Therefore, DSL enjoys a smaller generation error bound than supervised learning.

References

1. Ahmad, W. U., Chakraborty, S., Ray, B., & Chang, K.-W. (2020). A transformer-based approach for source code summarization. Preprint. arXiv:2005.00653.
2. Alon, U., Brody, S., Levy, O., & Yahav, E. (2018). code2seq: Generating sequences from structured representations of code. In *International Conference on Learning Representations*.
3. Bahdanau, D., Cho, K., & Bengio, Y. (2015). Neural machine translation by jointly learning to align and translate. In *Third International Conference on Learning Representations, ICLR 2015*.
4. Bajaj, P., Campos, D., Craswell, N., Deng, L., Gao, J., Liu, X., et al. (2016). MS MARCO: A human generated machine reading comprehension dataset. Preprint. arXiv:1611.09268.
5. Bartlett, P. L., & Mendelson, S. (2002). Rademacher and Gaussian complexities: Risk bounds and structural results. *Journal of Machine Learning Research, 3*(Nov), 463–482.
6. Burges, C. J. C. (1998). A tutorial on support vector machines for pattern recognition. *Data Mining and Knowledge Discovery, 2*(2), 121–167.
7. Cao, R., Zhu, S., Liu, C., Li, J., & Yu, K. (2019). Semantic parsing with dual learning. In *Proceedings of the 57th Annual Meeting of the Association for Computational Linguistics* (pp. 51–64).
8. Cui, S., Lian, R., Jiang, D., Song, Y., Bao, S., & Jiang, Y. (2019). Dal: Dual adversarial learning for dialogue generation. In *Proceedings of the Workshop on Methods for Optimizing and Evaluating Neural Language Generation* (pp. 11–20).
9. Dai, A. M., & Le, Q. V. (2015). Semi-supervised sequence learning. In *Advances in Neural Information Processing Systems* (pp. 3079–3087).
10. Fernandes, P., Allamanis, M., & Brockschmidt, M. (2018). Structured neural summarization. In *International Conference on Learning Representations*.
11. Fuglede, B., & Topsoe, F. (2004). Jensen–Shannon divergence and Hilbert space embedding. In *Proceedings of the International Symposium on Information Theory, 2004. ISIT 2004* (p. 31). Piscataway, NJ: IEEE.
12. Hayati, S. A., Olivier, R., Avvaru, P., Yin, P., Tomasic, A., & Neubig, G. (2018). Retrieval-based neural code generation. In *Proceedings of the 2018 Conference on Empirical Methods in Natural Language Processing* (pp. 925–930).
13. He, S., Han, C., Han, G., & Qin, J. (2020). Exploring duality in visual question-driven top-down saliency. *IEEE Transactions on Neural Networks and Learning Systems, 31*(7), 2672–2679.
14. He, K., Zhang, X., Ren, S., & Sun, J. (2016). Deep residual learning for image recognition. In *Proceedings of the IEEE Conference on Computer Vision and Pattern Recognition* (pp. 770–778).
15. Jean, S., Cho, K., Memisevic, R., & Bengio, Y. (2015). On using very large target vocabulary for neural machine translation. In *Proceedings of the 53rd Annual Meeting of the Association for Computational Linguistics and the 7th International Joint Conference on Natural Language Processing* (pp. 1–10).

16. Kingma, D. P., & Ba, J. (2014). Adam: A method for stochastic optimization. Preprint. arXiv:1412.6980.
17. Krizhevsky, A. (2009). Learning multiple layers of features from tiny images. Technique Report.
18. Ling, W., Blunsom, P., Grefenstette, E., Hermann, K. M., Kočiský, T., Wang, F., et al. (2016). Latent predictor networks for code generation. In *Proceedings of the 54th Annual Meeting of the Association for Computational Linguistics* (Vol. 1: Long Papers), pp. 599–609.
19. Maas, A. L., Daly, R. E., Pham, P. T., Huang, D., Ng, A. Y., & Potts, C. (2011). Learning word vectors for sentiment analysis. In *Proceedings of the 49th Annual Meeting of the Association for Computational Linguistics: Human Language Technologies* (Vol. 1, pp. 142–150). Stroudsburg, PA: Association for Computational Linguistics.
20. Manning, C. D., Manning, C. D., & Schütze, H. (1999). *Foundations of statistical natural language processing*. Cambridge, MA: MIT Press.
21. Mikolov, T., Karafiát, M., Burget, L., Černocký, J., & Khudanpur, S. (2010). Recurrent neural network based language model. In *Eleventh Annual Conference of the International Speech Communication Association.*
22. Mohri, M., Rostamizadeh, A., & Talwalkar, A. (2018). *Foundations of Machine Learning*. Cambridge, MA: MIT Press.
23. Novikova, J., Dušek, O., & Rieser, V. (2017). The e2e dataset: New challenges for end-to-end generation. In *Proceedings of the 18th Annual SIGdial Meeting on Discourse and Dialogue* (pp. 201–206).
24. Rajpurkar, P., Zhang, J., Lopyrev, K., & Liang, P. (2016). Squad: 100,000+ questions for machine comprehension of text. In *Proceedings of the 2016 Conference on Empirical Methods in Natural Language Processing* (pp. 2383–2392).
25. Salimans, T., Karpathy, A., Chen, X., & Kingma, D. P. (2017). Pixelcnn++: Improving the pixelcnn with discretized logistic mixture likelihood and other modifications. Preprint. arXiv:1701.05517.
26. Sennrich, R., Haddow, B., & Birch, A. (2016). Neural machine translation of rare words with subword units. In *Proceedings of the 54th Annual Meeting of the Association for Computational Linguistics* (Vol. 1: Long Papers, pp. 1715–1725).
27. Su, S.-Y., Huang, C.-W., & Chen, Y.-N. (2019). Dual supervised learning for natural language understanding and generation. In *Proceedings of the 57th Annual Meeting of the Association for Computational Linguistics* (pp. 5472–5477).
28. Sun, Y., Tang, D., Duan, N., Qin, T., Liu, S., Yan, Z., et al. (2020). Joint learning of question answering and question generation. *IEEE Transactions on Knowledge and Data Engineering, 32*(5), 971–982.
29. Sundermeyer, M., Schlüter, R., & Ney, H. (2012). LSTM neural networks for language modeling. In *Thirteenth Annual Conference of the International Speech Communication Association.*
30. Tibshirani, R. (1996). Regression shrinkage and selection via the lasso. *Journal of the Royal Statistical Society: Series B (Methodological), 58*(1), 267–288.
31. Vaswani, A., Shazeer, N., Parmar, N., Uszkoreit, J., Jones, L., Gomez, A. N., et al. (2017). Attention is all you need. In *Advances in Neural Information Processing Systems* (pp. 5998–6008).
32. Wei, B., Li, G., Xia, X., Fu, Z., & Jin, Z. (2019). Code generation as a dual task of code summarization. In *Advances in Neural Information Processing Systems* (pp. 6559–6569).
33. Xia, Y., Qin, T., Chen, W., Bian, J., Yu, N., & Liu, T.-Y. (2017). Dual supervised learning. In *Proceedings of the 34th International Conference on Machine Learning* (Vol. 70, pp. 3789–3798). JMLR.org
34. Yang, Y., Yih, W.-t., & Meek, C. (2015). WikiQA: A challenge dataset for open-domain question answering. In *Proceedings of the 2015 Conference on Empirical Methods in Natural Language Processing* (pp. 2013–2018).

35. Ye, W., Xie, R., Zhang, J., Hu, T., Wang, X., & Zhang, S. (2020). Leveraging code generation to improve code retrieval and summarization via dual learning. In *Proceedings of The Web Conference 2020* (pp. 2309–2319)
36. Yin, P., & Neubig, G. (2017). A syntactic neural model for general-purpose code generation. In *Proceedings of the 55th Annual Meeting of the Association for Computational Linguistics* (Vol. 1: Long Papers, pp. 440–450).
37. Zeiler, M. D. (2012). Adadelta: An adaptive learning rate method. Preprint. arXiv:1212.5701

Chapter 8
Dual Inference

So far in the previous chapters of this book, we have mainly discussed how to leverage structure duality to improve model training. When the primal and dual models have been well trained, the primal model is used to do inference for the primal task, and the dual model is used to do inference for the dual task. There is no interaction between the two models in the inference procedure.

Actually, structure duality can also be used to improve the inference procedure. Intuitively, we have high confidence to judge y is a good output for the input x in the primal task, if x is a good output for y in the dual task, and vice versa. Dual inference is a new formulation using both the pre-trained primal and dual models to conduct inference for each individual task.

8.1 General Formulation

In this section, we introduce the general formulation of dual inference.

First, let us have a look on how standard inference works in conventional supervised learning. Suppose a model parameterized by θ is trained by maximum likelihood estimation,

$$\max_{\theta} \sum_{(x,y)} \log P(y|x;\theta).$$

Naturally, in inference, given an input x, the model outputs a y that can maximize the conditional probability:

$$y = \arg\max_{y' \in \mathcal{Y}} P(y'|x;\theta).$$

T. Qin, *Dual Learning*, https://doi.org/10.1007/978-981-15-8884-6_8

Similarly, in dual learning works introduced in previous chapters, we conduct inference for the primal and dual tasks works as follows.

$$y = \arg\max_{y' \in \mathcal{Y}} P(y'|x; \theta_{XY})$$

$$x = \arg\max_{x' \in \mathcal{X}} P(x'|y; \theta_{YX})$$

According to the joint-probability principle (see Eqn. (7.1)), the conditional probability $P(y|x)$ can be computed using the primal model θ_{XY}, and also using the dual model θ_{YX}:

$$P(y|x) = \frac{P(x, y)}{P(x)} = \frac{P(y)P(x|y; \theta_{YX})}{P(x)}.$$

Similarly, we have

$$P(x|y) = \frac{P(x, y)}{P(y)} = \frac{P(x)P(y|x; \theta_{XY})}{P(Y)}.$$

Given that the conditional probability can be computed in two ways using the primal and dual models separately, it is natural to combine them together. Dual inference is such an inference framework that conducts inference for each individual task by combining together the conditional probabilities computed by the primal and dual models. For primal task,

$$P(y|x; \theta_{XY}, \theta_{YX}) = \alpha P(y|x; \theta_{XY}) + (1 - \alpha)\frac{P(y)P(x|y; \theta_{YX})}{P(x)},$$

and for dual task,

$$P(x|y; \theta_{XY}, \theta_{YX}) = \beta P(x|y; \theta_{YX}) + (1 - \beta)\frac{P(x)P(y|x; \theta_{XY})}{P(Y)},$$

where $\alpha \in [0, 1]$ and $\beta \in [0, 1]$ are hyper parameters controlling the trade-off between of the two terms and will be tuned and determined based on performance on a validation set.

While the above two equations are derived from the equation of joint probability for learning tasks with maximum likelihood estimation, Xia et al. [8] provide a more general formulation that works for any primal/dual tasks with well defined loss functions. We here describe the general formulation.

Let $f : \mathcal{X} \mapsto \mathcal{Y}$ denote the model for the primal task which is a mapping from space $\mathcal{X}$ to space $\mathcal{Y}$, and $g : \mathcal{Y} \mapsto \mathcal{X}$ denote the model for the dual task. Let $\ell_f(x, y)$ and $\ell_g(x, y)$ denote the loss functions for f and g respectively.

Xia et al. [8] introduce the following general formulation for dual inference,

$$f_{\text{dual}}(x) = \arg\min_{y' \in \mathcal{Y}} \{\alpha \ell_f(x, y') + (1 - \alpha)\ell_g(x, y')\},$$

$$g_{\text{dual}}(y) = \arg\min_{x' \in \mathcal{X}} \{\beta \ell_g(x', y) + (1 - \beta)\ell_f(x', y)\}.$$

Here we make some discussions about dual inference.

First, dual inference does not re-train or make any changes to the models of the primal and dual tasks. It only modifies the inference rules.

Second, most of inference rules currently widely-used in machine learning tasks can be described as

$$f(x) = \arg\min_{y' \in \mathcal{Y}} \ell_f(x, y')$$

for the primal task and

$$g(y) = \arg\min_{x' \in \mathcal{X}} \ell_g(x', y)$$

for the dual task, which can be viewed as extreme cases of dual inference by setting $\alpha = 1$ and $\beta = 1$. From this perspective, dual inference can be viewed as a more general inference framework.

Third, dual inference is conceptually related to model ensemble [5], which also employs multiple models for inference to achieve better accuracy. Dual inference can be viewed as a special kind of model ensemble. Different from most model ensemble methods, in which all models follow the same mapping direction and unidirectional inference is improved, dual inference involves two (primal and dual) models with opposite mapping directions and both directional inferences are boosted.

8.2 Applications

Thanks to its general formulation, dual inference can be applied to many different applications, as shown in [8]. We take neural machine translation as an example here.

The loss functions can be specialized as the negative log-likelihood in machine translation:

$$\begin{aligned} \ell_f(x, y) &= -\log P(y|x; f), \\ \ell_g(x, y) &= -\log P(x|y; g). \end{aligned} \tag{8.1}$$

Recall that in neural machine translation, the output space $\mathcal{Y}$ is huge and there are exponentially many possible y's for an input sentence x. Thus it is impossible to go through all possible y's to find the one with the minimal loss, and one relies on beam search to find a reasonably good y. Dual inference also relies on beam search to find a set of good candidates using the primal or dual model, then leverages both models to re-rank those candidates, and finally output the best one. The detailed process of dual inference for the primal task of neural machine translation is shown as follows:

1. Translate source x with beam search by model f and get K candidates $\{\hat{y}_i\}_{i\in[K]}$; (K is beam size)
2. Find the best candidate

$$i^* = \arg\min_{i\in[K]} \alpha\ell_f(x, \hat{y}_i) + (1-\alpha)\ell_g(x, \hat{y}_i),$$

   where ℓ_f and ℓ_g are defined in Eq. (8.1).
3. Return $\hat{y}_{i^*}$ as the translation of x.

Dual inference for the dual task of neural machine translation can be conducted in the same way. We omit details here.

In addition to neural machine translation, Xia et al. [8] also show that dual inference can improve inference accuracy for sentiment analysis (sentiment classification vs. sentence generation conditioned on a sentiment label) and image processing (image classification vs. image generation conditioned on a category label).

Duan et al. [3] study the problem of question generation and question answering. They show that a well trained question generation model can help question answering (more accurately, answer sentence selection) through dual inference on three benchmark datasets: MARCO [1], SQUAD [6], and WikiQA [9].

Zheng et al. [10] study the problem of quantile modeling, which involves two models: a Q model that predicts the corresponding value for an arbitrary quantile, and an F model that predicts the corresponding quantile for an arbitrary value. Clearly, the two models are in the dual form. Zhang et al. take this duality into account and show that joint training can be performed for the two models and dual inference achieves better performance for quantile regressions.

8.3 Theoretical Analysis

In standard supervised learning, a model is trained by minimizing some loss function over the training dataset, and then is used to make prediction for an unseen instance in the test set by minimizing the same loss function. The training procedure is consistent with the inference procedure. However, in dual inference, the training and inference procedures are inconsistent with each other. Taking the primal task as an example, only the primal model is trained, but both the primal and dual models are used in inference. One may concern that whether there still exist theoretical guarantees for the effectiveness of dual inference. To address this concern, in this section, we show that under mild assumptions, dual inference does have theoretical guarantees. In particular, we present a data-dependent bound for classification tasks with dual inference.

We make two assumptions for theoretical analysis:

1. $\mathcal{Y} = \{1, 2, \cdots, c\}$ where $c \geq 2$. That is, we assume the primal task is classification, such as sentiment classification in sentiment analysis and image classification in image processing. Note that this assumption limits the application scope of the theoretical results we obtain in this section.
2. For any $x \in \mathcal{X}$ and $y \in \mathcal{Y}$, $\ell_f(x, y) \in [0, 1]$ and $\ell_g(x, y) \in [0, 1]$. This assumption is not difficult to satisfy. For example, if the loss functions ℓ_f and ℓ_g are bounded, we can scale them into [0, 1].

Let φ_f denote $1 - \ell_f$ and let φ_g denote $1 - \ell_g$. We define

$$\varphi = \alpha\varphi_f + (1 - \alpha)\varphi_g.$$

We further define the margin

$$\rho(x, y) = \varphi(x, y) - \max_{y' \neq y} \varphi(x, y').$$

Thus, φ misclassifies (x, y) iff $\rho(x, y) \leq 0$.

Let $\mathcal{S} = ((x_1, y_1), \ldots, (x_m, y_m))$ denote a training set of m samples drawn i.i.d from some unknown distribution D over $\mathcal{X} \times \mathcal{Y}$. For any $\rho > 0$, the generalization error $R(\varphi)$ and its empirical margin error $\hat{R}_{S,\rho}$ are defined as follows.

$$R(\varphi) = \mathbb{E}_{(x,y)\sim D}[1\{\rho_\varphi(x, y) \leq 0\}]$$

$$\hat{R}_{S,\rho} = \frac{1}{m}\sum_{i=1}^{m}[1\{\rho_\varphi(x_i, y_i) \leq \rho\}]$$

For any family of hypothesis G mapping $\mathcal{X} \times \mathcal{Y}$ to $\mathbb{R}$, we define $\Pi_1(G)$ as

$$\Pi_1(G) = \{x \mapsto h(x, y) : y \in \mathcal{Y}, h \in G\}.$$

Let $\mathcal{H}_f$ and $\mathcal{H}_g$ denote the two hypothesis spaces of φ_f and φ_g. Let $\mathfrak{R}_m(\cdot)$ denote the Rademacher complexity [2]. We leverage the Theorem 1 in [4], further optimize it under our settings, and obtain the following theorem:

Theorem 8.1 *Fix* $\rho > 0$*, for any* $\delta > 0$*, with probability at least* $1 - \delta$ *over the choice of a training set* S *of size* m *drawn i.i.d. according to* D*, the following inequality holds:*

$$R(\varphi) \leq \hat{R}_{S,\rho}(\varphi) + \frac{8c}{\rho}\Big(\alpha\mathfrak{R}_m(\Pi_1(\mathcal{H}_f)) + (1-\alpha)\mathfrak{R}_m(\Pi_1(\mathcal{H}_g))\Big)$$

$$+\frac{1}{\rho}\sqrt{\frac{2}{m}} + \sqrt{\frac{1}{2m}\log\Big(\lceil\frac{4}{\rho^2}\log(\frac{mc^2\rho^2}{2})\rceil + 1\Big) + \frac{1}{2m}\log\frac{1}{\delta}}.$$

Theorem 8.1 tells that the generalization error of dual inference is related to the Rademacher complexities of both $\mathcal{H}_f$ and $\mathcal{H}_g$. The hyper-parameter α for dual inference is explicitly considered in the bound.

According to [4], we know that if using the primal model φ_f only, we have

$$R(\varphi_f) \leq \hat{R}_{S,\rho}(\varphi_f) + \frac{4c}{\rho}\mathfrak{R}_m(\Pi_1(\mathcal{H}_f)) + \sqrt{\frac{1}{2m}\log\frac{1}{\delta}}. \tag{8.2}$$

Since dual learning and dual inference are mostly using neural network models, according to [7], we have that the $\mathfrak{R}_m(\Pi_1(\cdot))$'s in Theorem 8.1 and (8.2) are of order $O(\sqrt{1/m})$ under regular assumptions. Thus, the generalization error bounds of standard inference and dual inference are $O(\sqrt{1/m})$ and $O(\sqrt{\log\log(m)/m})$ respectively. This shows that dual inference has comparable generalization error bound with that of standard inference. That is, the inconsistency between the training and inference procedures does not hurt the theoretical performance of dual inference. Actually, empirical results [3, 8, 10] demonstrate that by leveraging the primal and dual models, dual inference performs much better than standard inference using only one model.

References

1. Bajaj, P., Campos, D., Craswell, N., Deng, L., Gao, J., Liu, X., et al. (2016) MS MARCO: A human generated machine reading comprehension dataset. Preprint. arXiv:1611.09268.
2. Bartlett, P. L., & Mendelson, S. (2002). Rademacher and Gaussian complexities: Risk bounds and structural results. *Journal of Machine Learning Research, 3*(Nov), 463–482.
3. Duan, N., Tang, D., Chen, P., & Zhou, M. (2017). Question generation for question answering. In *Proceedings of the 2017 Conference on Empirical Methods in Natural Language Processing* (pp. 866–874)
4. Kuznetsov, V., Mohri, M., & Syed, U. (2014). Multi-class deep boosting. In *Advances in Neural Information Processing Systems* (pp. 2501–2509)

5. Opitz, D., & Maclin, R. (1999). Popular ensemble methods: An empirical study. *Journal of Artificial Intelligence Research, 11*, 169–198.
6. Rajpurkar, P., Zhang, J., Lopyrev, K., & Liang, P. (2016). Squad: 100,000+ questions for machine comprehension of text. In *Proceedings of the 2016 Conference on Empirical Methods in Natural Language Processing* (pp. 2383–2392).
7. Sun, S., Chen, W., Wang, L., Liu, X., & Liu, T.-Y. (2016). On the depth of deep neural networks: A theoretical view. In *Thirtieth AAAI Conference on Artificial Intelligence*.
8. Xia, Y., Bian, J., Qin, T., Yu, N., & Liu, T.-Y. (2017). Dual inference for machine learning. In *Proceedings of the 26th International Joint Conference on Artificial Intelligence* (pp. 3112–3118).
9. Yang, Y., Yih, W.-t., & Meek, C. (2015). WikiQA: A challenge dataset for open-domain question answering. In *Proceedings of the 2015 Conference on Empirical Methods in Natural Language Processing* (pp. 2013–2018).
10. Zhang, F., Fan, X., Xu, H., Zhou, P., He, Y., & Liu, J. (2019). Regression via arbitrary quantile modeling. Preprint. arXiv:1911.05441.

Chapter 9
Marginal Probability Based Dual Semi-Supervised Learning

We have introduced dual learning from unlabeled data (including both the semi-supervised setting and the unsupervised setting) based on the dual reconstruction principle in Chaps. 4–6. We have also introduced the joint-probability principle, which enables dual learning in the supervised setting and in inference. In this chapter we introduce a new perspective of dual learning from unlabeled data, in which marginal probability of unlabeled data is the focus and the dual model is used to efficiently approximate the marginal probability of a sample.

This chapter is organized as follows. We first introduce the key, how to efficiently approximate marginal distributions using the dual model, in Sect. 9.1, then introduce the works that constrains the marginal probability of unlabeled data in Sect. 9.2 and maximizes the likelihood of unlabeled data in Sect. 9.3, and finally make some discussions in Sect. 9.4.

9.1 Efficient Approximation of Marginal Probability

The key challenge of marginal probability based dual semi-supervised learning is how to calculate the marginal probability ($P(x)$ or $P(y)$) efficiently, and the solution is to leverage the dual model to compute a sample weight and then efficiently approximate the marginal probability of a sample through importance sampling. In this section, we take neural machine translation (NMT) as an example to introduce the details of the challenge and the solution.

According to the law of total probability, the marginal probability $P(y)$ of a sentence y in language Y can be computed using the conditional probability $P(y|x)$:

$$P(y) = \sum_{x \in \mathcal{X}} P(y|x)P(x).$$

T. Qin, *Dual Learning*, https://doi.org/10.1007/978-981-15-8884-6_9

As will be seen in Sects. 9.2 and 9.3, this marginal probability can be utilized in different ways to learn from unlabeled data. All those ways face the same challenge: how to efficiently compute the marginal probability efficiently?

The naïve computation by summarizing over all the space $\mathcal{X}$ looks simple but computationally unaffordable because there are exponentially many candidate sentences in the space $\mathcal{X}$.

Another straightforward approach is to do Monte Carlo sampling in the space $\mathcal{X}$ and use the average of the samples to approximate the expectation:

$$\begin{aligned}\sum_{x\in\mathcal{X}} P(y|x)P(x) &= \mathbb{E}_{x\sim P(x)}P(y|x)\\ &\approx \frac{1}{K}\sum_{i=1}^{K} P(y|x^{(i)}), \qquad x^{(i)} \sim P(x).\end{aligned} \tag{9.1}$$

That is, given a target-language sentence $y \in \mathcal{Y}$, one samples K source sentences $x^{(i)}$ according to the marginal distribution $P(x)$, and then computes the average conditional probability over the K samples. However, if we estimate the expectation term by direct Monte Carlo sampling from distribution $P(x)$, we cannot get a good approximation of $\sum_{x\in\mathcal{X}} P(y|x)P(x)$ with a reasonably sized K. Intuitively, given a sentence y in the target language, when we sample a sentence x in the source language from the marginal distribution $P(x)$, which is a source-side language model, it is almost impossible that x is exactly or close to the translation of y. In other words, the sampled sentence x from $P(x)$ in the source language is usually irrelevant to the given sentence y in the target language. Therefore, almost all the source-language sentences sampled from distribution $P(x)$ will result in $P(y|x)$ very close to zero.

In order to well approximate $\sum_{x\in\mathcal{X}} P(y|x)P(x)$, we should draw samples with relatively large $P(y|x)$, i.e., making sampled sentences x relevant to the given sentence y. Clearly, if we have a well-trained target-to-source language translation model, we can obtain several candidate translation x's of y using such a translation model, which is exactly the dual translation model.

Since we sample from the conditional distribution $P(x|y)$, which is characterized by the dual model, instead of the marginal distribution $P(x)$, we need to adjust our estimate of $\sum_{x\in\mathcal{X}} P(y|x)P(x)$ as follows:

$$\sum_{x\in\mathcal{X}} P(y|x)P(x) = \sum_{x\in\mathcal{X}} \frac{P(y|x)P(x)}{P(x|y)} P(x|y) = \mathbb{E}_{x\sim P(x|y)} \frac{P(y|x)P(x)}{P(x|y)}.$$

That is, by making a multiplicative adjustment to $P(y|x)$ we compensate for sampling from $P(x|y)$ instead of $P(x)$. This procedure is exactly the technique of *importance sampling* [1, 3, 4]. Then, we arrive at a new estimation of $\sum_{x \in \mathcal{X}} P(y|x)P(x)$ by importance sampling using the dual model $P(x|y)$:

$$\sum_{x \in \mathcal{X}} P(y|x)P(x) \approx \frac{1}{K} \sum_{i=1}^{K} \frac{P(y|x^{(i)})P(x^{(i)})}{P(x^{(i)}|y)}, \qquad x^{(i)} \sim P(x|y), \tag{9.2}$$

where K is the sample size. This new estimation is a much better approximation of $P(y)$ than that in Eq. (9.1).

9.2 Marginal Probability as a Constraint

Wang et al. [5] study semi-supervised neural machine translation and leverage monolingual data from the probability perspective. The key idea is to take the marginal probability of unlabeled data as a constraint to regularize the training of an NMT model.

We first give some notations. Let X and Y denote the source and target languages respectively, $P(y|x;\theta_{XY})$ denote the probability of translating a source-language sentence x to a target-language sentence y characterized by a source-to-target translation (the primal) model with parameters θ_{XY}, $\mathcal{B}$ denote the corpus of bilingual sentence pairs, and $\mathcal{M}_Y$ denote the monolingual corpus of the target language Y.

In supervised learning for neural machine translation, the translation model θ_{XY} is learned by maximizing the likelihood of the paired data

$$l(\theta_{XY}) = \frac{1}{|\mathcal{B}|} \sum_{(x,y)\in\mathcal{B}} \log P(y|x;\theta_{XY}). \tag{9.3}$$

Given a target-language sentence y, according to the law of total probability, we have

$$P(y) = \sum_{x \in \mathcal{X}} P(y|x)P(x),$$

where $\mathcal{X}$ is the sentence space of language X. If the translation model θ_{XY} is perfect, we should have

$$P(y) = \sum_{x \in \mathcal{X}} P(y|x;\theta_{XY})P(x). \tag{9.4}$$

However, since θ_{XY} is empirically trained through likelihood maximization on the bilingual data, there is no guarantee that Eq. (9.4) will hold for a general target-language sentence y. Therefore, Wang et al. [5] propose to enhance model training by forcing all the sentences in $\mathcal{M}_Y$ to satisfy the probability relation in

Eq. (9.4). Mathematically, the training is reformulated as the following constrained optimization problem:

$$\max \frac{1}{|\mathcal{B}|} \sum_{(x,y)\in\mathcal{B}} \log P(y|x;\theta_{XY})$$

$$s.t. P(y) = \sum_{x\in\mathcal{X}} P(y|x;\theta_{XY})P(x), \forall y \in \mathcal{M}_Y.$$

Following the common practice in constrained optimization, Wang et al. convert the constraint in the above optimization problem into the following regularization term

$$\left[\log P(y) - \log \sum_{x\in\mathcal{X}} P(y|x;\theta_{XY})P(x)\right]^2.$$

Adding this term to the likelihood objective, we get the following loss function to minimize for semi-supervised machine translation:

$$l(\theta_{XY}) = -\frac{1}{|\mathcal{B}|} \sum_{(x,y)\in\mathcal{B}} \log P(y|x;\theta_{XY}) + \lambda \frac{1}{|\mathcal{M}_Y|} \sum_{y\in\mathcal{M}_Y} \left[\log P(y) - \log \sum_{x\in\mathcal{X}} P(y|x;\theta_{XY})P(x)\right]^2, \tag{9.5}$$

where λ is a hyper-parameter controlling the trade-off between the likelihood and the regularization term.

Note that the marginal distributions $P(x)$ and $P(y)$ are not available. Similar to that in Chap. 7.3, Wang et al. use the empirical distributions $\hat{P}(x)$ and $\hat{P}(y)$ as their surrogates, which are modeled by well-trained language models. One can use large scale monolingual corpora to train language models for the source and target languages.

Now the only blocking issue to minimize the loss in Eq. (9.5) is how to efficiently compute the expectation $\sum_{x\in\mathcal{X}} P(y|x;\theta_{XY})P(x)$, which is exactly the focus of Sect. 9.1, as shown in Eq. (9.2). By substituting Eq. (9.2) into Eq. (9.5), we get the final training objective:

$$l(\theta_{XY}) = -\frac{1}{|\mathcal{B}|} \sum_{(x,y)\in\mathcal{B}} \log P(y|x;\theta_{XY}) + \lambda \frac{1}{|\mathcal{M}_Y|} \sum_{y\in\mathcal{M}_Y} \left[\log P(y) - \log \frac{1}{K} \sum_{i=1}^{K} \frac{P(y|x^{(i)};\theta_{XY})P(x^{(i)})}{P(x^{(i)}|y;\theta_{YX})}\right]^2, \tag{9.6}$$

where $x^{(i)}$ is sampled from $P(x|y;\theta_{YX})$ and θ_{YX} is a pre-trained dual model that translates a target-language sentence to a source-language sentence.

The experimental results in [5] show that:

- Maximum likelihood training plus the marginal probability constraint leads to significant accuracy improvement.
- Importance sampling with a larger sampling size K leads to higher translation accuracy, but at the cost of more computations.
- A small K (e.g., 2 or 3) arrives at a good trade-off between accuracy improvement and computational cost.
- The better translation quality the dual model θ_{YX} is of, the larger improvement one can get for the primal model θ_{XY}.

9.3 Likelihood Maximization for Unlabeled Data

Different from [5], in which monolingual data is used as a constraint that connects the translation model to the marginal probability of a target-language sentence, Wang et al. [6] propose a new way to utilize the target-language monolingual data by directly maximizing the log likelihood of a target-language sentence.

Mathematically, for a sentence $y \in \mathcal{M}_Y$, its likelihood is

$$\log P(y) = \log \sum_{x \in \mathcal{X}} P(y|x)P(x).$$

Combining with the likelihood objective in Eq. (9.3), we get the following negative log likelihood:

$$\begin{aligned} l(\theta_{XY}) = &-\frac{1}{|\mathcal{B}|}\sum_{(x,y)\in\mathcal{B}} \log P(y|x;\theta_{XY}) \\ &-\lambda\frac{1}{|\mathcal{M}_Y|}\sum_{y\in\mathcal{M}_Y} \log \sum_{x\in\mathcal{X}} P(y|x;\theta_{XY})P(x), \end{aligned} \tag{9.7}$$

where λ is a hyper-parameter controlling the trade-off between the likelihood on the bilingual data and the likelihood on the monolingual data.

Again, to make the above loss practically computable, we need to substitute Eq. (9.2) into Eq. (9.7) and get the following approximated loss function

$$
\begin{aligned}
l(\theta_{XY}) \approx & -\frac{1}{|\mathcal{B}|} \sum_{(x,y)\in\mathcal{B}} \log P(y|x;\theta_{XY}) \\
& -\lambda \frac{1}{|\mathcal{M}_Y|} \sum_{y\in\mathcal{M}_Y} \log \frac{1}{K} \sum_{i=1}^{K} \frac{P(y|x^{(i)};\theta_{XY})P(x^{(i)})}{P(x^{(i)}|y;\theta_{YX})},
\end{aligned}
\tag{9.8}
$$

where $x^{(i)}$ is sampled from $P(x|y;\theta_{YX})$ and θ_{YX} is a pre-trained dual model.

The overall training procedure of dual semi-supervised machine translation is shown in Algorithm 1. Note that this algorithm also works for the loss defined in Eq. (9.6) in Sect. 9.2, and we only need to replace the loss $l(\theta_{XY})$ in Line 6 and 7 of the algorithm.

Algorithm 1 Dual semi-supervised machine translation with monolingual data likelihood maximization

Require: Bilingual corpus $\mathcal{B}$, monolingual corpus $\mathcal{M}_Y$, pre-trained dual translation model $P(x|y;\theta_{YX})$, empirical marginal distributions $\hat{P}(x)$ and $\hat{P}(y)$, hyper-parameters λ, sample size K.

1: Randomly initialize the translation model $P(y|x;\theta_{XY})$.
2: Train the translation model $P(y|x;\theta_{XY})$ by maximizing the bilingual data likelihood in Eq. (9.3).
3: For each sentence y in $\mathcal{M}_Y$, sample K sentences $x^{(1)}, \ldots x^{(K)}$ according to the dual translation model $P(x|y;\theta_{YX})$;
4: **repeat**
5: Sample a mini-batch of monolingual sentences from $\mathcal{M}_Y$ and a mini-batch of bilingual sentence pairs from $\mathcal{B}$;
6: Calculate the loss $l(\theta_{XY})$ defined in Eq. (9.8) using the sampled two mini-batches of data.
7: Update the parameters θ_{XY} with learning rate γ:

$$\theta_{XY} \leftarrow \theta_{XY} - \gamma \nabla_{\theta_{XY}} l(\theta_{XY}).$$

8: **until** the model θ_{XY} converges

9.4 Discussions

The dual semi-supervised learning algorithms introduced in previous sections can be extended from different aspects.

Combination of the Objectives The two objectives defined in Eq. (9.6) and (9.8) can be combined. Wang et al. [6] show that the combination of two objectives leads to further accuracy improvement.

Joint Training of the Primal and Dual Models Both the algorithms proposed in [5, 6] fix the dual model (the target-to-source language translation) and use it to help the training of the primal model (the source-to-target language translation). This can be viewed as a special kind of transfer learning, transferring the knowledge from the dual translation task to help the learning of the primal translation task, and thus Wang et al. position their work as dual transfer learning in [5]. Actually it is easy to extend the algorithms in [5, 6] so that the primal and dual models can both be updated and mutually boosted. We show a direct extension as follows.

1. We define the bilingual data likelihood as

$$l(\theta_{YX}) = \frac{1}{|\mathcal{B}|} \sum_{(x,y)\in\mathcal{B}} \log P(x|y;\theta_{YX}). \tag{9.9}$$

2. Following the derivations in Sect. 9.2, we can get the following negative log likelihood enhanced by constraining the marginal probability of monolingual source-language data:

$$\begin{aligned} l(\theta_{YX}) \approx &-\frac{1}{|\mathcal{B}|} \sum_{(x,y)\in\mathcal{B}} \log P(x|y;\theta_{YX})) \\ &+\lambda \frac{1}{|\mathcal{M}_X|} \sum_{x\in\mathcal{M}_X} \left[\log P(x) - \log \frac{1}{K} \sum_{i=1}^{K} \frac{P(x|y^{(i)};\theta_{YX})P(y^{(i)})}{P(y^{(i)}|x;\theta_{XY})}\right]^2, \end{aligned} \tag{9.10}$$

   where $y^{(i)}$ is sampled from $P(y|x;\theta_{XY})$.
3. Following the derivations in Sect. 9.3, we can get the following combined negative log likelihood of the bilingual data and the monolingual source-language data:

$$\begin{aligned} l(\theta_{XY}) \approx &-\frac{1}{|\mathcal{B}|} \sum_{(x,y)\in\mathcal{B}} \log P(x|y;\theta_{YX})) \\ &-\lambda \frac{1}{|\mathcal{M}_X|} \sum_{x\in\mathcal{M}_X} \log \frac{1}{K} \sum_{i=1}^{K} \frac{P(x|y^{(i)};\theta_{YX})P(y^{(i)})}{P(y^{(i)}|x;\theta_{XY})}, \end{aligned} \tag{9.11}$$

   where $y^{(i)}$ is sampled from $P(y|x;\theta_{XY})$.
4. Now we can design an algorithm (as below) that leverages both source-language and target-language monolingual corpora to boost both the primal and dual models.

More Applications While we focus on the application of machine translation in this chapter, it is not difficult to apply the ideas and extend the algorithms introduced

Algorithm 2 Marginal probability enhanced dual semi-supervised machine translation

Require: Bilingual corpus $\mathcal{B}$, monolingual source-language corpus $\mathcal{M}_X$, monolingual target-language corpus $\mathcal{M}_Y$, empirical marginal distributions $\hat{P}(x)$ and $\hat{P}(y)$, hyper-parameters λ, and sample size K.

1: Randomly initialize the translation models $P(y|x;\theta_{XY})$ and $P(x|y;\theta_{YX})$.
2: Train the translation model $P(y|x;\theta_{XY})$ by maximizing the bilingual data likelihood in Eq. (9.3).
3: Train the translation model $P(x|y;\theta_{YX})$ by maximizing the bilingual data likelihood in Eq. (9.9).
4: **repeat**
5: Sample a mini-batch of monolingual sentences M_X from $\mathcal{M}_X$, a mini-batch of monolingual sentences M_Y from $\mathcal{M}_Y$, and a mini-batch of bilingual sentence pairs B from $\mathcal{B}$.
6: Calculate the gradient $\nabla_{\theta_{XY}} l(\theta_{XY})$ of the primal model with respect to the loss $l(\theta_{XY})$ defined in Eq. (9.6) or Eq. (9.8) using the data M_Y and B.
7: Calculate the gradient $\nabla_{\theta_{YX}} l(\theta_{YX})$ of the dual model with respect to the loss $l(\theta_{YX})$ defined in Eq. (9.10) or Eq. (9.11) using the data M_X and B.
8: Update the two models with learning rate γ:

$$\theta_{XY} \leftarrow \theta_{XY} - \gamma \nabla_{\theta_{XY}} l(\theta_{XY}),$$

$$\theta_{YX} \leftarrow \theta_{YX} - \gamma \nabla_{\theta_{YX}} l(\theta_{YX}).$$

9: **until** the models θ_{XY} and θ_{YX} converge

here to other applications such as question answering and generation, text summarization, code summarization and generation, etc. We expect there will be more dual semi-supervised learning algorithms designed for different applications.

Combination of Different Principles The works introduced in this chapter leverage unlabeled data based on the principle of marginal probability. We have introduced the dual reconstruction principle [2, 7, 8] in Chap. 4.2, which is also for learning with unlabeled data. It is not difficult to combine the two principles to further enhance the learning from unlabeled data.

References

1. Cochran, W. G. (2007). *Sampling techniques*. John Wiley & Sons.
2. He, D., Xia, Y., Qin, T., Wang, L., Yu, N., Liu, T.-Y., et al. (2016). Dual learning for machine translation. In *Advances in Neural Information Processing Systems* (pp. 820–828).
3. Hesterberg, T. (1995). Weighted average importance sampling and defensive mixture distributions. *Technometrics, 37*(2), 185–194.
4. Neal, R. M. (2001). Annealed importance sampling. *Statistics and computing, 11*(2), 125–139.
5. Wang, Y., Xia, Y., Zhao, L., Bian, J., Qin, T., Liu, G., et al. (2018). Dual transfer learning for neural machine translation with marginal distribution regularization. In *Thirty-Second AAAI Conference on Artificial Intelligence*.

6. Wang, Y., Xia, Y., Zhao, L., Bian, J., Qin, T., Chen, E., et al. (2019). Semi-supervised neural machine translation via marginal distribution estimation. *IEEE/ACM Transactions on Audio, Speech, and Language Processing, 27*(10), 1564–1576.
7. Yi, Z., Zhang, H., Tan, P., & Gong, M. (2017). Dualgan: Unsupervised dual learning for image-to-image translation. In *Proceedings of the IEEE International Conference on Computer Vision* (pp. 2849–2857).
8. Zhu, J.-Y., Park, T., Isola, P., & Efros, A. A. (2017). Unpaired image-to-image translation using cycle-consistent adversarial networks. In *Proceedings of the IEEE International Conference on Computer Vision* (pp. 2223–2232).

Part IV
Advanced Topics

This part focuses on several advanced topics, including theoretical understandings of dual reconstruction and connections between dual learning and other learning paradigms.

Chapter 10
Understanding Dual Reconstruction

10.1 Overview

In this chapter, we focus on theoretical understandings of the principle of dual reconstruction, the foundation of the algorithms introduced in Chaps. 4–6. There are two reasons motivating us to theoretically understand the principle.

First, while dual learning algorithms have enjoyed great success in unsupervised settings [2, 9, 11, 12, 19, 22], this is neither natural nor intuitive. Take image to image translations in Fig. 5.5 as an example. Without any supervisions, why DiscoGAN can discover a meaningful relationship between images of cars and images of faces or between images of chairs and images of cars? Note that without supervisions, there exist unlimited numbers of possible mappings from a domain to another domain. Actually, it is surprising that DiscoGAN can find such a semantic mapping from unlimited possibilities.

Second, while dual learning algorithms for supervised settings [13, 18] and the inference settings [17], which are based on the joint-probability principle, have been studied empirically and theoretically simultaneously, unsupervised and semi-supervised algorithms built on the dual reconstruction principle are only studied empirically [2, 8, 9, 11, 12, 19, 22] without theoretical understandings. This might imply that theoretical analysis of the principle is challenging.

Given the great success of those algorithms built on the principle of dual reconstruction, it is naturally to ask why they work and in which situations they will fail. Galanti et al. [6] and Zhao et al. [20] study the principle in different settings. We introduce them in this chapter.

T. Qin, *Dual Learning*, https://doi.org/10.1007/978-981-15-8884-6_10

10.2 Understanding Dual Reconstruction in Unsupervised Settings

Galanti et al. [6] consider unsupervised settings and analyze from the perspective of the complexity of mapping functions, i.e., the complexity of the primal and dual models.

10.2.1 A Formulation of Dual Unsupervised Mapping

Unsupervised dual learning aims to learn a primal model/function f_{XY} that maps an instance in domain $\mathcal{X}$ to an instance in domain $\mathcal{Y}$ and a dual model/function f_{YX} that maps an instance in domain $\mathcal{Y}$ to an instance in domain $\mathcal{X}$, given two unlabeled datasets $\{x_i\}$ i.i.d sampled from domain $\mathcal{X}$ according to distribution D_X and $\{y_j\}$ i.i.d sampled from domain $\mathcal{Y}$ according to distribution D_Y.

To formally analyze unsupervised dual learning algorithms [2, 9, 11, 19, 22], Galanti et al. [6] introduce the third domain $\mathcal{Z}$ associated with the distribution D_Z and assume that there exist two mapping functions $f_{ZX} : \mathcal{Z} \to \mathcal{X}$ and $f_{ZY} : \mathcal{Z} \to \mathcal{Y}$ satisfying

$$D_X = f_{ZX} \circ D_Z$$

and

$$D_Y = f_{ZY} \circ D_Z,$$

where $f_{ZX} \circ D_Z$ denotes the distribution of $f_{ZX}(z)$ with $z \sim D_Z$. Note that this assumption does not hold naturally and certain conditions are required. For example, the domain $\mathcal{Z}$ should not be too simpler than $\mathcal{X}$ and $\mathcal{Y}$. Fortunately, such a condition is easy to be satisfied in most applications studied for unsupervised dual learning, including machine translation and image to image translation. Actually, for image to image translation, the three domains are the same image space containing images of the same size.

Galanti et al. further assume that both f_{ZX} and f_{ZY} are invertible, which is empirically justified by the recent success of supervised pre-image computation methods [4]. Similarly, in dual unsupervised learning, we assume that the mapping functions f_{XY} and f_{YX} are invertible.

We denote by

$$f_{XY} = f_{ZY} \circ f_{ZX}^{-1},$$

which means that to map domain $\mathcal{X}$ to domain $\mathcal{Y}$, we first map it to $\mathcal{Z}$ and then to $\mathcal{Y}$. This indirect mapping implies the underlying semantic assumption: the domain

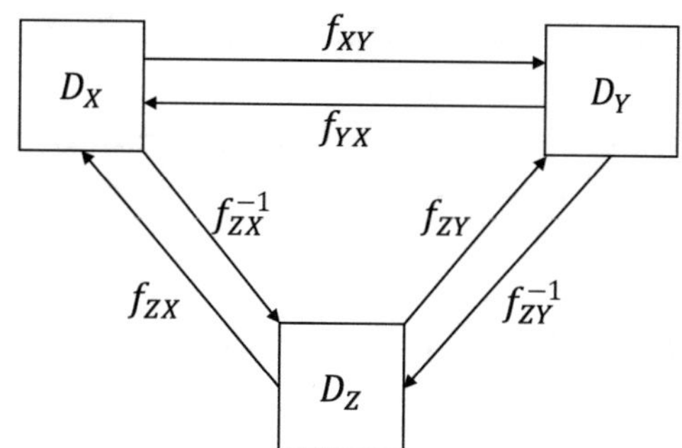

Fig. 10.1 The distributions of three domains and the mapping functions between them

$\mathcal{Z}$ is a semantic space shared by $\mathcal{X}$ and $\mathcal{Y}$ and thus the mapping between the two domains can be bridged by the semantic space, as shown in Fig. 10.1.

Galanti et al. [6] define the "alignment" problem as fitting a function $h \in \mathcal{H}$, for some hypothesis class $\mathcal{H}$ that is closest to f_{XY},

$$\inf_{h \in \mathcal{H}} R_{D_X}[h, f_{XY}],$$

where

$$R_D[f_1, f_2] = \mathbb{E}_{x \sim D} \ell(f_1(x), f_2(x))$$

measures the distance between two functions f_1 and f_2 with respect to a distribution D using a loss function $\ell : \mathbb{R} \times \mathbb{R} \to \mathbb{R}$. Note that this is unsupervised alignment because no paired data between the two domains $\mathcal{X}$ and $\mathcal{Y}$ are used.

To ensure that we can learn a good primal model h and a good dual model h', dual unsupervised learning algorithms are built on the dual reconstruction principle and the adversarial training principle with the following training objective:

$$\begin{aligned} \inf_{h,h' \in \mathcal{H}} & \operatorname{disc}_{\mathcal{C}}(h \circ D_X, D_Y) + \operatorname{disc}_{\mathcal{C}}(h' \circ D_Y, D_X) \\ & + R_{D_X}[h' \circ h, \mathrm{Id}_X] + R_{D_Y}[h \circ h', \mathrm{Id}_Y] \end{aligned} \tag{10.1}$$

where Id_X and Id_Y denote identity mappings in domain X and Y separately, $\operatorname{disc}_{\mathcal{C}}(D_1, D_2)$ denotes the discrepancy[1] between distributions D_1 and D_2 measured by a function class with complexity at most C and is usually implemented with a GAN [7]. The first term in Eq. (10.1) ensures that the data distribution induced from h that maps from domain X to domain Y is close to the original data distribution in domain Y. The second term is the analog term for the mapping in the reverse direction. The last two terms are the dual reconstruction error to ensure that a sample can be reconstructed through the forward and backward mappings.

[1]Formal definition of $\operatorname{disc}_{\mathcal{C}}(D_1, D_2)$ can be found at [6].

10.2.2 Issues and the Simplicity Hypothesis

On the one hand, the dual reconstruction errors expressed as the last two terms in Eq. (10.1) are simple and elegant and do not require additional supervision for training; on the other hand, they have a potential issue as pointed out by Galanti et al. [6]:

Proposition 10.1 *If there exist two mapping functions h and h' leading to zero reconstruction error, i.e.,*

$$h \circ h' = \mathrm{Id}_X, \quad h \circ h' = \mathrm{Id}_Y,$$

there are many pairs of such kind of mappings with zero reconstruction error.

This proposition is not difficult to prove. For every invertible permutation[2] Π of the samples in domain Y, one has

$$\begin{aligned}(h' \circ \Pi^{-1}) \circ (\Pi \circ h) = h \circ h' = \mathrm{Id}_X, \text{ and} \\ (\Pi \circ h) \circ (h' \circ \Pi^{-1}) = \Pi \circ (h \circ h') \circ \Pi^{-1} = \Pi \circ \mathrm{Id}_Y \circ \Pi^{-1} = \mathrm{Id}_Y.\end{aligned} \tag{10.2}$$

Therefore, every pair h and h' that ensures dual reconstruction gives rise to many possible solutions of the form $\tilde{h} = h \circ \Pi$ and $\tilde{h}' = \Pi^{-1} \circ h'$. If Π happens to satisfy $D_X(x) \approx D_X(\Pi(x))$, then the discrepancy terms in Eq. (10.1) also remain largely unchanged. Here is a simple example for this bad situation.

Example 10.2 Suppose we want to translate English words "one", "two" and "three" to Chinese words[3] "1", "2" and "3". The correct translation should be "one"↔"1", "two"↔"2" and "three" ↔"3". Unfortunately, there are six translations (e.g., the translation of "one"↔"2", "two"↔"3" and "three" ↔"1") leading to zero dual reconstruction error. Furthermore, if "one", "two" and "three" are of the same probability in English and "1", "2" and "3" are of the same probability in Chinese, those six translations will lead to the same discrepancy loss. That is, the six translations are of the same training error, while only one of them is semantically correct.

As shown by the above example and Eq. (10.2), the unsupervised learning of mappings between two domains without unpaired data is ill posed, and the dual reconstruction principle by itself (even if enhanced by adversarial training) cannot explain the recent success of dual unsupervised learning. However, the algorithms in [9, 19, 22] enjoy empirical success, even though there exist a large number of

[2] Assume that both domains contain finite number of elements, which is the case for image to image translation with fixed image size and machine translation with maximal sentence length constraint.

[3] For simplicity, here we use Arabic numbers to denote Chinese words.

alternative mappings that satisfy the constraints of Eq. (10.1) and are semantically incorrect. Why?

To answer this question, Galanti et al. [6] make the following simplicity hypothesis:

Hypothesis 10.3 (The Simplicity Hypothesis) The small-discrepancy mapping of the lowest complexity approximates the alignment of the target function.

The intuition of this hypothesis is that the correct forward and dual mappings should be the simplest pair among all mapping pairs that minimize the objective in Eq. (10.1). It coincide with Occam's Razor, which also prefers simplicity in problem solving.

Galanti et al. further hypothesize that in those dual unsupervised algorithms [2, 9, 11, 19, 22], goldilock architectures are selected, which are complex enough to allow small discrepancies but not over complex to support mappings that are not minimal in complexity. Therefore, one of the minimal-complexity low-discrepancy mappings is learned.

Based on these hypothesises, Galanti et al. make some predictions.

Prediction 10.4 For unsupervised learning of mappings between domains that share common characteristics, if the network is small enough, the GAN constraint in the target domain is sufficient to obtain a semantically aligned mapping.

Remark At the first glance, this prediction seems to eliminate the necessity of the dual reconstruction principle and the joint training of the primal and dual models, which contradicts with the observations and claims in [9, 19, 22]. Actually, as shown in [6], without the dual reconstruction principle, the GAN constraint works well for highly related domains sharing common characteristics such as male $\leftrightarrow$ female face conversion, black $\leftrightarrow$ blond hair conversion, with $\leftrightarrow$ without eyeglasses conversion, and edges $\leftrightarrow$ shoes conversion, but does not learn meaningful mappings for domains sharing less common characteristics such as handbag $\leftrightarrow$ shoes conversion. Note that previous works [9, 19, 22] show that the GAN constraint together with the dual reconstruction principle can indeed learn good mappings for those less related domains (e.g., handbag $\leftrightarrow$ shoes conversion), and can discover better mappings for highly related domains than the GAN constraint only. Thus, the dual reconstruction principle plays an important role in learning unsupervised mappings between two domains, which cannot be explained by the simplicity hypothesis.

Prediction 10.5 For unsupervised learning of a mapping between two domains, the complexity of the network needs to be carefully adjusted.

This is a natural extension of Prediction 10.4 for the case when small networks are not expressive enough to learn semantically reasonable mappings.

If the simplicity hypothesis is correct, we need to find a pair of primal and dual mappings with low complexity and small discrepancies. However, deeper architectures can lead to smaller discrepancies. Thus, we need to trade off the complexity of the networks and the distribution discrepancy. For this purpose, Galanti et al. propose to find a function h of a non-minimal complexity k_2 that minimizes the following objective function

$$\min_{h \text{ s.t } C(h)=k_2} \left\{ \text{disc}(h \circ D_X, D_Y) + \lambda \inf_{g \text{ s.t } C(g)=k_1} R_{D_X}[h, g] \right\}, \tag{10.3}$$

where k_1 is the minimal complexity for mapping from domain X to domain Y with small discrepancy. In other words, we want to find a function h that is of low distribution discrepancy and close to the minimal-complexity mapping g with small discrepancy.

10.2.3 Minimal Complexity

To formally study their hypotheses, Galanti et al. [6] introduce the concept of minimal complexity.

Let us consider image translation tasks and focus on functions with the following form

$$f := F[W_{n+1}, \ldots, W_1] = W_{n+1} \circ \sigma \circ \ldots \circ \sigma \circ W_2 \circ \sigma \circ W_1, \tag{10.4}$$

where $W_1, \ldots, W_{n+1}$ are invertible linear transformations from $\mathbb{R}^d$ to itself and σ is a non-linear element-wise activation function. Galanti et al. mainly focus on σ that is Leaky ReLU with parameter $0 < a \neq 1$.

Definition 10.6 The complexity of a function f, denoted by $C(f)$, is defined as the minimal number n such that there are invertible linear transformations $W_1, \ldots, W_{n+1}$ that satisfy $f = F[W_{n+1}, \ldots, W_1]$.

Intuitively, $C(f)$ is the depth of the shallowest neural network, if there are multiple such networks, that can implement the function f. That is, we use the number of layers of a network as a proxy for the Kolmogorov complexity of functions, which is simpler than Kolmogorov complexity when we are studying functions that can be computed by feedforward neural networks.

Note that Galanti et al. [6] consider the complexity of individual functions. There are many complexity concepts defined for function classes, including the VC-dimension [15] and Rademacher complexity [3]. How to leverage those complexity concepts to analyze dual learning algorithms (especially the dual reconstruction principle) is not explored yet, which is a good future research direction for interested readers.

Definition 10.7 (Density Preserving Mapping) A ϵ_0-density preserving mapping over a domain $X = (\mathcal{X}, D_X)$ (or an ϵ_0-DPM for short) is a function f such that

$$\text{disc}(f \circ D_X, D_X) \leq \epsilon_0 \tag{10.5}$$

Let $\mathrm{DPM}_{\epsilon_0}(X;k) := \{f \mid \mathrm{disc}(f \circ D_X, D_X) \le \epsilon_0 \text{ and } C(f) = k\}$ denote the set of all ϵ_0-DPMs of complexity k.

Galanti et al. further make the following prediction/assumption.

Prediction 10.8 The number of DPMs of low complexity is small.

This prediction means that DPMs are expected to be rare in real-world domains and will be used to upper bound the number of minimal low-discrepancy mappings in Theorem 10.10. As neural network based mappings/functions are continuous and uncountable, we need to carefully characterize the number of minimal low-discrepancy mappings. For this purpose, we give some further definitions.

Based on Definition 10.6, we define ϵ-similarity in terms of density preserving as follows.

Definition 10.9 Two functions f and g are ϵ-similar with respect to distribution D, denoted as $f \overset{D}{\underset{\epsilon_0}{\sim}} g$, if $C(f) = C(g)$ and there are minimal decompositions: $f = F[W_{n+1}, \ldots, W_1]$ and $g = F[V_{n+1}, \ldots, V_1]$ such that: $\forall i \in [n+1]$: $\mathrm{disc}(F[W_i, \ldots, W_1] \circ D, F[V_i, \ldots, V_1] \circ D) \le \epsilon_0$.

According to the above definition, two functions are ϵ-similar with respect to D if (1) they are of the same minimal complexity, and (2) for every step of their processing, the activations of the matching functions preserve density (with respect to D).

There are many different ways to partition a set/space $\mathcal{U}$ of functions into disjoint subsets such that in each subset, any two functions are ϵ-similar. Let $\mathrm{N}(\mathcal{U}, \sim_{\mathcal{U}})$ denote the minimal number of subsets required in order to cover the entire space $\mathcal{U}$, where $\sim_{\mathcal{U}}$ is the similarity relation. This can be viewed as a generalized version of covering numbers [21].

Let $C^{\epsilon_0}_{X,Y}$ denote the minimal complexity of the networks needed in order to achieve discrepancy smaller than ϵ_0 for mapping the distribution D_X to the distribution D_Y. Let $H_{\epsilon_0}(X,Y)$ denote the set of minimal complexity mappings, i.e., mappings of complexity $C^{\epsilon_0}_{X,Y}$ that achieve ϵ_0 discrepancy:

$$H_{\epsilon_0}(X,Y) := \left\{h \mid C(h) \le C^{\epsilon_0}_{X,Y} \text{ and } \mathrm{disc}(h \circ D_X, D_Y) \le \epsilon_0\right\}.$$

Galanti et al. prove the following theorem, which indicates that the covering number of this set is similar to the covering number of the DPMs. Therefore, if Prediction 10.8 holds, the number of minimal low-discrepancy mappings is small.

Theorem 10.10 *Let σ be a Leaky ReLU with parameter $0 < a \ne 1$ and assume identifiability. For $\epsilon_0 > 0$ and $0 < \epsilon_2 < \epsilon_1$, we have*

$$\mathrm{N}\left(H_{\epsilon_0}(X,Y), \overset{D_X}{\underset{\epsilon_1}{\sim}}\right) \le \min \begin{cases} \mathrm{N}\left(\mathrm{DPM}_{\epsilon_0}\left(X; 2C^{\epsilon_0}_{X,Y}\right), \overset{D_X}{\underset{\epsilon_2}{\sim}}\right) \\ \mathrm{N}\left(\mathrm{DPM}_{\epsilon_0}\left(Y; 2C^{\epsilon_0}_{X,Y}\right), \overset{D_Y}{\underset{\epsilon_2}{\sim}}\right). \end{cases} \tag{10.6}$$

Note that the theorem assumes identifiability, which is about the uniqueness of neural networks up to invariants, and is still an open problem for general networks. Progress has been made for specific network architectures. [5] proves the identifiability of feedforward networks with tanh activations, [1, 10, 14, 16] prove the uniqueness (up to some invariants) for neural networks with only one hidden layer and various activation functions, and Galanti et al. [6] prove the identifiability of leaky ReLU networks with one hidden layer.

10.3 Understanding Dual Reconstruction in Semi-Supervised Settings

Proposition 10.2 and Example 10.2 show that for the pair of oracle mappings (i.e., the forward mapping and backward mapping), there exist many mapping pairs (e.g., by permuting the oracle mappings) that are semantically incorrect but lead to zero or small dual reconstruction error. Galanti et al. [6] explain the success of dual learning algorithms based on the simplicity hypothesis. A limitation of this work is that the simplicity hypothesis itself is not theoretically proved; even worse, in many applications like machine translation and speech processing, neural networks used for mappings are complex and may violate the simplicity hypothesis.

To better understand and explain the success of dual learning algorithms, Zhao et al. [20] analyze the dual reconstruction principle in semi-supervised settings and focus on the task of machine translation [8]. They first show that dual learning can improve the original mappings for the translations between two languages (Sect. 10.3.2) and then extend the results to more languages (Sect. 10.3.3).

10.3.1 Algorithm and Notations

The dual reconstruction based semi-supervised dual learning algorithm (see Algorithm 1) for machine translation starts from two vanilla translators T_{12} and T_{21} trained from bilingual data $\mathcal{B}_{12}$ and gets two improved translators T_{12}^d and T_{21}^d, where the superscript d stands for dual learning. We want to understand why after dual learning T_{12}^d and T_{21}^d have better accuracy than the vanilla ones T_{12} and T_{21}. To conduct formal analysis, we first give some notations.

Let $L_i, i \in \{1, 2, \cdots, k\}$ denote the i-th language space, composed of sentences in the language. Let D_i denote the distribution of sentences in L_i, i.e., $\Pr(x) = D_i(x), \forall x \in L_i$. As the vocabulary of a language is of limited size and the length of a sentence is also limited, we assume there are a finite number of sentences in each language space. Furthermore, let $c(x)$ denote the cluster of sentences sharing the same semantic meanings as a sentence x.

Algorithm 1 Semi-supervised dual learning

Require: Bilingual data $\mathcal{B}_{12}$ for two languages L_1 and L_2, monolingual data $\mathcal{M}_1$ for L_1 and $\mathcal{M}_2$ for L_2.

1: Train vanilla translators T_{12} and T_{21} for $L_1 \to L_2$ and $L_2 \to L_1$ translation respectively using bilingual data.

2: Continue training both translators through dual learning by minimizing the negative log likelihood on the bilingual data $\mathcal{B}_{12}$ and the dual reconstruction error on monolingual data $\mathcal{M}_1$ and $\mathcal{M}_2$.

3: Output the final translators T_{12}^d and T_{21}^d.

Let T_{ij}^* denote the oracle translator that maps a cluster or a sentence in the cluster in language L_i to the correct cluster in language L_j. Let T_{ij} denote a vanilla translator that translates from a sentence in L_i to one in L_j. The desired mapping is $T_{ij}(x_i) \in T_{ij}^*(x_i)$. Let p_{ij} denote the expected accuracy of the translator T_{ij}, i.e., the probability of correctly translating a sentence randomly sampled from L_i according to D_i. Formally,

$$p_{ij} = \Pr_{x \sim D_i}(T_{ij}(x) \in T_{ij}^*(x)) = \sum_{x \in L_i} D_i(x)\mathbb{1}[T_{ij}(x) \in T_{ij}^*(x)],$$

where $\mathbb{1}[]$ is the indicator function. For simplicity, we will omit the subscript $x \sim D_i$ when the context is clear and there is no confusion. It is easy to see that

$$\Pr(T_{ij}(x) \notin T_{ij}^*(x)) = 1 - p_{ij}.$$

Let D_j' denote the distribution of sentences in language L_j derived from $T_{ij}(x)$ where $x \sim D_i$. We define

$$p_{ji}^r = \Pr_{x \sim D_j'}(T_{ji}(x) \in T_{ji}^*(x)) = \sum_{x \in L_j} D_j'(x)\mathbb{1}[T_{ji}(x) \in T_{ji}^*(x)].$$

The superscript r means "reconstruction". The difference between p_{ji}^r and p_{ji} lies in the distributions of samples in space L_j.

10.3.2 Translation Between Two Languages

For each sentence $x \in L_1$, let $y_{12}(x)$ and $y_{21}(x)$ denote whether $T_{12}(x)$ and $T_{21}(T_{12}(x))$ produce correct translations:

$$y_{12}(x) = \begin{cases} 1, & \text{if } T_{12}(x) \in T_{12}^*(x), \\ 0, & \text{otherwise,} \end{cases}$$

and

$$y_{21}(x) = \begin{cases} 1, & \text{if } T_{21}(T_{12}(x)) \in T_{21}^{*}(T_{12}(x)), \\ 0, & \text{otherwise.} \end{cases}$$

Then by definition, we have

$$p_{12} = \Pr_{x \sim D_1}(y_{12}(x) = 1) \tag{10.7}$$

$$p_{21}^{r} = \Pr_{x \sim D_1}(y_{21}(x) = 1). \tag{10.8}$$

When there is no confusion, we omit the subscript $x \sim D_1$ for simplicity.

In order to analyze dual learning, we consider the joint distribution of y_{12} and y_{21}. We introduce λ to model the dependence of y_{12} and y_{21}:

$$\Pr(y_{12} = 1, y_{21} = 1) = p_{12}p_{21}^{r} + \lambda. \tag{10.9}$$

Then we have

$$\Pr(y_{12} = 1, y_{21} = 0) = p_{12}(1 - p_{21}^{r}) - \lambda,$$
$$\Pr(y_{12} = 0, y_{21} = 1) = (1 - p_{12})p_{21}^{r} - \lambda,$$
$$\Pr(y_{12} = 0, y_{21} = 0) = (1 - p_{12})(1 - p_{21}^{r}) + \lambda,$$

using Eq. (10.7) and (10.8).

Because all these probabilities are non-negative, we have

$$-\min\{p_{12}p_{21}^{r}, (1 - p_{12})(1 - p_{21}^{r})\} \le \lambda \le \min\{p_{12}, p_{21}^{r}\}. \tag{10.10}$$

Although this range of λ is not tight, it is sufficient for our analysis.

The probability of the alignment issue, which means for some $x \in L_1$, $T_{21}(T_{12}(x)) \in c(x)$ and $y_{12}(x) = y_{21}(x) = 0$, is part of $\Pr(y_{12}(x) = 0, y_{21}(x) = 0)$. We use δ to model how likely this issue occurs:

$$p_{\text{align}} = \delta((1 - p_{12})(1 - p_{21}^{r}) + \lambda), \tag{10.11}$$

where $0 \le \delta \le 1$.

For translators T_{12}^{d} and T_{21}^{d} obtained from dual learning, we define $y_{12}^{d}()$ and $y_{21}^{d}()$ similarly:

$$y_{12}^{d}(x) = \begin{cases} 1, & \text{if } T_{12}^{d}(x) \in T_{12}^{*}(x), \\ 0, & \text{otherwise,} \end{cases}$$

and

$$y_{21}^d(x) = \begin{cases} 1, & \text{if } T_{21}^d(T_{12}^d(x)) \in T_{21}^*(T_{12}^d(x)), \\ 0, & \text{otherwise.} \end{cases}$$

We are interested in the accuracy of T_{12}^d:

$$p_{12}^d = \Pr_{x \sim D_1}(y_{12}^d(x) = 1).$$

To bridge the vanilla translators and dual translators, Zhao et al. make the following assumption, which says if a sample in L_1 is successfully reconstructed by vanilla translators, it is also successfully reconstructed by dual translators.

Assumption 10.11 *For any $x \in L_1$, if $T_{21}(T_{12}(x)) \in c(x)$, then $T_{21}^d(T_{12}^d(x)) \in c(x)$ holds.*

For simplicity, we denote this case as Case 1 and the remaining cases as Case 2. Formally, for any $x \in L_1$,

Case 1: $T_{21}(T_{12}(x)) \in c(x)$;
Case 2: $T_{21}(T_{12}(x)) \notin c(x)$.

For any given $x \in L_1$ which falls in Case 2, we define

$$\alpha = \Pr(T_{12}^d(x) \in T_{12}^*(x), T_{21}^d(T_{12}^d(x)) \in c(x)|\text{Case 2}) \tag{10.12}$$

$$\beta = \Pr(T_{12}^d(x) \notin T_{12}^*(x), T_{21}^d(T_{12}^d(x)) \in c(x)|\text{Case 2}) \tag{10.13}$$

$$\gamma = \Pr(T_{21}^d(T_{12}^d(x)) \notin c(x)|\text{Case 2}), \tag{10.14}$$

where "Case 2" denotes the condition $T_{21}(T_{12}(x)) \notin c(x)$. Here α can be viewed as the probability of correcting the wrong translations by dual learning, β the probability of the occurrence of the alignment problem under Case 2, and γ the probability of nonzero reconstruction error. γ models the imperfectness of dual learning, which should be zero in the ideal case. It is easy to see $\alpha + \beta + \gamma = 1$.

The following theorem characterizes the accuracy of the translator T_{12}^d after dual learning.

Theorem 10.12 *Under Assumption 10.11, for two language L_1 and L_2, the accuracy of dual learning outcome T_{12}^d is*

$$p_{12}^d = (1-\alpha)(p_{12}p_{21}^r + \lambda) + \alpha\delta(p_{12} + p_{21}^r - p_{12}p_{21}^r - \lambda) + \alpha(1-\delta),$$

where λ, δ, α are defined in Eq. (10.9), (10.11) *and* (10.14).

Proof Consider a random sample x and the translation from $x \in L_1$ to L_2. Before dual learning, the accuracy is p_{12}. We analyze the aforementioned two cases.

Case 1. $T_{21}(T_{12}(x)) \in c(x)$.

This case consists of two sub cases:

- Case 1.1: $T_{12}(x) \in T_{12}^*(x)$;
- Case 1.2: $T_{12}(x) \notin T_{12}^*(x)$.

Although Case 1.2 is not desired, dual learning does not detect it. From Eq. (10.9) and (10.11), the probabilities of the Case 1.1 and Case 1.2 are

$$\Pr(\text{Case 1.1}) = \Pr(y_{12} = y_{21} = 1) = p_{12}p_{21}^r + \lambda,$$

$$\Pr(\text{Case 1.2}) = p_{\text{align}} = \delta((1 - p_{12})(1 - p_{21}^r) + \lambda).$$

Case 2. $T_{21}(T_{12}(x)) \notin c(x)$.

Dual learning will train the translators so that this case is minimized. The probability of this case is simply the complement of Case 1:

$$\begin{aligned}\Pr(\text{Case 2}) &= 1 - (p_{12}p_{21}^r + \lambda) - \delta((1 - p_{12})(1 - p_{21}^r) + \lambda)\\ &= 1 - \delta - (1 + \delta)(p_{12}p_{21}^r + \lambda) + \delta(p_{12} + p_{21}^r).\end{aligned}$$

After dual learning, Case 2 is redistributed to Case 1.1 and Case 1.2, with probabilities α and β respectively. So we have

$$\begin{aligned}&\Pr(T_{12}^d(x) \in T_{12}^*(x), T_{21}^d(T_{12}^d(x)) \in c(x))\\ =&p_{12}p_{21}^r + \lambda + \alpha \Pr(\text{Case 2})\\ =&(1 - \alpha)(p_{12}p_{21}^r + \lambda) + \alpha\delta(p_{12} + p_{21}^r - p_{12}p_{21}^r - \lambda) + \alpha(1 - \delta),\end{aligned}$$

which is the accuracy of the translator T_{12}^d after dual learning.

We can draw several conclusions from the theorem.

First, we check relation between the translators before and after dual learning. Observing that $1 - \alpha \geq 0$ and $p_{12} + p_{21}^r - p_{12}p_{21}^r - \lambda \geq 0$ (due to Eq. (10.10)) and $1 - \delta \geq 0$, the accuracy of the translator after dual learning is positively correlated to the accuracy the vanilla translators of both directions before dual learning. The larger the p_{12} or p_{21}^r is, the higher accuracy of T_{12}^d after dual learning can achieve.

Second, we investigate how α and δ impact p_{12}^d. We have

$$p_{12}^d = \alpha(1 - \delta - (1 + \delta)(p_{12}p_{21}^r + \lambda) + \delta(p_{12} + p_{21}^r)) + p_{12}p_{21}^r + \lambda$$

by reorganization. So a larger α is desirable, which is consistent with our intuition. Also, p_{12}^d can be reorganized as

$$-\alpha\delta((1 - p_{12})(1 - p_{21}^r) + \lambda) + \alpha + (1 - \alpha)(p_{12}p_{21}^r + \lambda),$$

which means a small δ is desirable.

Third, we consider the case where the probabilities of redistribution to α case and β case are proportional to Pr(Case 1.1) and Pr(Case 1.2). Formally,

$$\frac{\alpha}{\beta} = \frac{\Pr(T_{12}(x) \in T^*_{12}(x), T_{21}(T_{12}(x)) \in c(x))}{\Pr(T_{12}(x) \notin T^*_{12}(x), T_{21}(T_{12}(x)) \in c(x))}$$
$$= \frac{p_{12}p^r_{21} + \lambda}{\delta((1 - p_{12})(1 - p^r_{21}) + \lambda)}.$$

Then we have

$$p^d_{12} = \frac{(p_{12}p^r_{21} + \lambda)(1 - \gamma(1 - p_{12}p^r_{21} - \lambda - \delta((1 - p_{12})(1 - p^r_{21}) + \lambda)))}{p_{12}p^r_{21} + \lambda + \delta((1 - p_{12})(1 - p^r_{21}) + \lambda)}$$
$$= \frac{(p_{12}p^r_{21} + \lambda)(1 - \Gamma)}{p_{12}p^r_{21} + \lambda + \delta((1 - p_{12})(1 - p^r_{21}) + \lambda)}, \tag{10.15}$$

where $\Gamma = \gamma(1 - p_{12}p^r_{21} - \lambda - \delta((1 - p_{12})(1 - p^r_{21}) + \lambda))$. To compare p^d_{12} with the accuracy of the vanilla translator, we compute the difference

$$p^d_{12} - p_{12} = p_{12}\left(\frac{(p^r_{21} + \lambda/p_{12})(1 - \Gamma)}{p_{12}p^r_{21} + \lambda + +\delta((1 - p_{12})(1 - p^r_{21}) + \lambda)} - 1\right)$$
$$= p_{12}\left(\frac{p^r_{21} + \lambda/p_{12}}{p_{12}p^r_{21} + \lambda + \delta((1 - p_{12})(1 - p^r_{21}) + \lambda)} - 1 - \Gamma\Delta\right)$$
$$= p_{12}\left(\frac{((1 + \delta)p^r_{21} - \delta)(1 - p_{12}) + \lambda(1/p_{12} - 1 + \delta)}{p_{12}p^r_{21} + \lambda + \delta((1 - p_{12})(1 - p^r_{21}) + \lambda)} - \Gamma\Delta\right),$$

where $\Delta = \frac{p^r_{21}+\lambda/p_{12}}{p_{12}p^r_{21}+\lambda+\delta((1-p_{12})(1-p^r_{21})+\lambda)}$.

Ideally, we have $\gamma = 0$, which means $\Gamma = 0$. If $p^r_{21} > \frac{\delta}{1+\delta}$, the outcome of dual learning is better than the vanilla translator. This condition is very mild because δ is small in general. The expression with the Γ factor is negative, which is consistent with the intuition that γ should be minimized.

10.3.3 *Extension: Multi-domain Dual Learning*

Theorem 10.12 shows that p_{ij} and p^r_{ji} play positive roles in improving p^d_{ij} under mild assumptions. A natural question is whether larger improvement can be achieved by exploiting more languages. Therefore, Zhao et al. [20] extend dual learning to multi-domain dual learning, by leveraging more language domains.

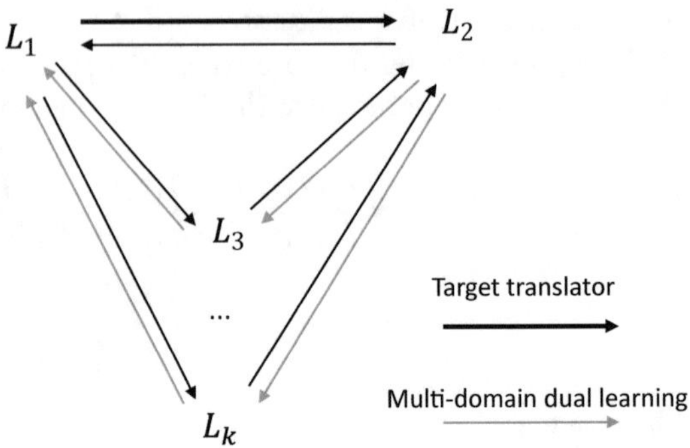

Fig. 10.2 Multi-domain dual learning. $L_1 \to L_2$ is the translation task of our focus, $L_2 \to L_1 \to L_2$ is the two-domain dual loop, and $L_2 \to L_3 \to L_1 \to L_2$ is the multi-domain dual loop

The basic idea of multi-domain dual learning is illustrated in Fig. 10.2, in which L_1 and L_2 denote the source language and target language , and $L_3, \cdots, L_k$ denote other helper languages. Multi-domain dual learning first trains translators $L_1 \leftrightarrow L_2$, $L_1 \leftrightarrow L_k$ and $L_2 \leftrightarrow L_k$ where $k \geq 3$ by standard dual learning. Then, it enforces a sentence from L_2 can be reconstructed through translations $L_2 \to L_k \to L_1 \to L_2$, and a sentence from L_1 can be reconstructed through $L_1 \to L_2 \to L_k \to L_1$). That is, multi-domain dual learning introduces another constraint, where the translation $L_1 \to L_2$ could leverage the information pivoted by domain L_k. The algorithm is formally shown as Algorithm 2, where θ_{ij} denotes the parameters of translator T_{ij}.

Algorithm 2 Multi-domain dual learning in semi-supervised settings

Require: Samples from $L_1 \ldots L_k$, initial translators T_{12}, T_{21} and T_{1i}, T_{i1} $\forall i = 3, \ldots, K$; learning rates η;

1: Train each of T_{12}, T_{21} and T_{1i}, T_{i1} $\forall i = 3, \ldots, k$ by dual learning with both bilingual data and monolingual data;

2: **repeat**

3: Randomly sample a k from $\{3, 4, \cdots, K\}$; randomly sample one $x^{(1)} \in L_1$ and one $x^{(2)} \in L_2$;

4: Generate $\tilde{x}^{(2)}$ by $T_{k2}(T_{1k}(x^{(1)}))$ and generate $\tilde{x}^{(1)}$ by $T_{k1}(T_{2k}(x^{(2)}))$;

5: Update the parameters of T_{12} and T_{21}, denoted as θ_{12} and θ_{21}, as follows:

$$\theta_{12} \leftarrow \theta_{12} + \eta \nabla_{\theta_{12}} \ln \Pr(x^{(2)} | \tilde{x}^{(1)}; \theta_{12});$$
$$\theta_{21} \leftarrow \theta_{21} + \eta \nabla_{\theta_{21}} \ln \Pr(x^{(1)} | \tilde{x}^{(2)}; \theta_{21}); \tag{10.16}$$

6: **until** convergence

Zhao et al. [20] theoretically analyze multi-domain dual learning and prove that it is better than standard dual learning under mild assumptions taking three languages as example. We omit the details here.

References

1. Albertini, F., Sontag, E. D., & Maillot, V. (1993). Uniqueness of weights for neural networks. *Artificial Neural Networks for Speech and Vision*, 115–125.
2. Artetxe, M., Labaka, G., Agirre, E., & Cho, K. (2018). Unsupervised neural machine translation. In *6th International Conference on Learning Representations.*
3. Bartlett, P. L., & Mendelson, S. (2002). Rademacher and gaussian complexities: Risk bounds and structural results. *Journal of Machine Learning Research, 3*(Nov), 463–482.
4. Dosovitskiy, A., & Brox, T. (2016). Inverting visual representations with convolutional networks. In *Proceedings of the IEEE Conference on Computer Vision and Pattern Recognition* (pp. 4829–4837).
5. Fefferman, C., & Markel, S. (1994). Recovering a feed-forward net from its output. In *Advances in Neural Information Processing Systems* (pp. 335–342).
6. Galanti, T., Wolf, L., & Benaim, S. (2018). The role of minimal complexity functions in unsupervised learning of semantic mappings. In *ICLR 2018: International Conference on Learning Representations 2018.*
7. Goodfellow, I., Pouget-Abadie, J., Mirza, M., Xu, B., Warde-Farley, D., Ozair, S., et al. (2014). Generative adversarial nets. In *Advances in Neural Information Processing Systems* (pp. 2672–2680).
8. He, D., Xia, Y., Qin, T., Wang, L., Yu, N., Liu, T.-Y., et al. (2016). Dual learning for machine translation. In *Advances in Neural Information Processing Systems* (pp. 820–828).
9. Kim, T., Cha, M., Kim, H., Lee, J. K., & Kim, J. (2017). Learning to discover cross-domain relations with generative adversarial networks. In *Proceedings of the 34th International Conference on Machine Learning-Volume 70* (pp. 1857–1865). JMLR.org.
10. Kurková, V., & Kainen, P. C. (2014). Comparing fixed and variable-width gaussian networks. *Neural Networks, 57*, 23–28.
11. Lample, G., Conneau, A., Denoyer, L., & Ranzato, M. (2018). Unsupervised machine translation using monolingual corpora only. In *6th International Conference on Learning Representations, ICLR 2018.*
12. Lin, J., Xia, Y., Qin, T., Chen, Z., & Liu, T.-Y. (2018). Conditional image-to-image translation. In *Proceedings of the IEEE Conference on Computer Vision and Pattern Recognition* (pp. 5524–5532).
13. Sun, Y., Tang, D., Duan, N., Qin, T., Liu, S., Yan, Z., et al. (2019). Joint learning of question answering and question generation. *IEEE Transactions on Knowledge and Data Engineering.*
14. Sussmann, H. J. (1992). Uniqueness of the weights for minimal feedforward nets with a given input-output map. *Neural Networks, 5*(4), 589–593.
15. Vapnik, V. N., & Chervonenkis, A. Ya. (1971). On the uniform convergence of relative frequencies of events to their probabilities. *Theory of Probability & Its Applications, 16*(2), 264–280.
16. Williamson, R. C., & Helmke, U. (1995). Existence and uniqueness results for neural network approximations. *IEEE Transactions on Neural Networks, 6*(1), 2–13.
17. Xia, Y., Bian, J., Qin, T., Yu, N., & Liu, T.-Y. (2017). Dual inference for machine learning. In *Proceedings of the 26th International Joint Conference on Artificial Intelligence* (pp. 3112–3118).
18. Xia, Y,. Qin, T., Chen, W., Bian, J., Yu, N., & Liu, T.-Y. (2017). Dual supervised learning. In *Proceedings of the 34th International Conference on Machine Learning-Volume 70* (pp. 3789–3798). JMLR.org.
19. Yi, Z., Zhang, H., Tan, P., & Gong, M. (2017). Dualgan: Unsupervised dual learning for image-to-image translation. In *Proceedings of the IEEE International Conference on Computer Vision* (pp. 2849–2857).
20. Zhao, Z., Xia, Y., Qin, T., Xia, L., & Liu, T.-Y. (2020). Dual learning: Theoretical study and an algorithmic extension.

21. Zhou, D.-X. (2002). The covering number in learning theory. *Journal of Complexity, 18*(3), 739–767.
22. Zhu, J.-Y., Park, T., Isola, P., & Efros, A. A. (2017). Unpaired image-to-image translation using cycle-consistent adversarial networks. In *Proceedings of the IEEE International Conference on Computer Vision* (pp. 2223–2232).

Chapter 11
Connections to Other Learning Paradigms

As we have discussed throughout this book, the key of dual learning is to leverage the structural duality between machine learning tasks to boost learning algorithms. When dual learning is studied in different settings, it may look similar to other related learning algorithms or paradigms. In this chapter, we discuss how dual learning is related to and different from some those learning algorithms and paradigms, including co-training, multi-task learning, GANs and autoencoders, Bayesian Ying-Yang learning, etc.

11.1 Dual Semi-Supervised Learning and Co-training

Co-training [1, 13] is a semi-supervised learning algorithm used to learn from a small amount of labeled data and a large amount of unlabeled data. It requires two views of the data and assumes that each data sample is described using two disjoint feature sets that provide different, complementary information about the instance:

- The two views are conditionally independent, which means that the features of data samples can be partitioned into two disjoint subsets and the two feature sets of each instance are conditionally independent given the class label.
- Each view is sufficient to make predictions, which means that the class label of an instance can be accurately predicted from each view (i.e., each feature subset) alone.

Co-training first learns a separate classifier for each view using all the labeled training examples. The most confident predictions of each classifier on the unlabeled data are then used to iteratively construct additional pseudo labeled training data to boost the other classifier.

Both dual semi-supervised learning and co-training are semi-supervised learning algorithms and focus on learning from labeled and unlabeled data. They are similar

T. Qin, *Dual Learning*, https://doi.org/10.1007/978-981-15-8884-6_11

in the sense that both involve two models and the two models are mutually boosted. They differ in a few aspects.

First, the most basic difference between dual semi-supervised learning and co-training is that the former one employs two heterogeneous models while the latter one employs two homogeneous models: in dual semi-supervised learning, the primal model maps from space $\mathcal{X}$ to $\mathcal{Y}$ and the dual model maps from space $\mathcal{Y}$ to $\mathcal{X}$; in co-training both the two classifiers maps from space $\mathcal{X}$ to $\mathcal{Y}$.

Second, their application scopes are different. Co-training is mainly for classification problems. In addition to classification, dual semi-supervised learning can also be applied to more complex problems, including (1) sequence generation problems such machine translation [5, 21], speech synthesis and recognition [15, 27], question answering and generation, and (2) image and video generation problems [8, 11, 28, 33].

Third, co-training is mainly based on conventional "shallow" machine learning models, since it was proposed before the renaissance of neural networks. Dual semi-supervised learning is based on deep neural networks, because dual learning was originally proposed to address the challenge of learning from limited labeled data for deep models.

Fourth, co-training makes strong assumptions about the conditional independence of the feature subsets and their expressiveness for predictions. Krogel and Scheffer [10] show that co-training is only beneficial if the data sets used in classification are independent. Dual semi-supervised learning does not rely on this kind of assumptions.

Last but not least, the effectiveness of co-training requires some subtle conditions.

- Co-training can only work if one of the classifiers correctly labels a piece of unlabeled data that the other classifier previously mis-classified. Unfortunately, when disagreements happen between the two classifiers, it is difficult to determine which classifier makes correct predictions for an unlabeled data sample.
- If both classifiers agree on all the unlabeled data, i.e. they are not independent, labeling the data does not create much new information. When applied to problems in functional genomics [10], co-training worsened the results as the dependence of the classifiers was greater than 60%.

In dual semi-supervised learning, even if both the primal and dual models make wrong predictions, it still provides effective feedback signal through the dual reconstruction error, which can be used to improve the two models. Furthermore, since the two models are heterogeneous, it will never happen that they agree with each other on the unlabeled data. Actually, the primal model leverages unlabeled data in space $\mathcal{X}$ and the dual model in space $\mathcal{Y}$. We do not need to worry about the independence of the two models.

11.2 Dual Learning and Multitask Learning

Multitask learning [2] is a sub branch of machine learning which conducts joint training for multiple learning tasks to exploit commonalities and differences across tasks. Compared with training the models separately, multitask learning can result in improved learning efficiency and prediction accuracy for the task-specific models. It has been explored in different applications such as natural language processing [3, 12] and computer vision [30] and studied with different model structures such as conventional shallow models [4] and deep neural networks [17].

Since dual learning (including dual semi-supervised learning, dual unsupervised learning and dual supervised learning) jointly trains the primal and dual models, it is a special case of multitask learning. The key difference between dual learning and general multitask learning is that the two tasks in dual learning are of structural duality, while the multiple tasks in multitask learning usually share the same input space but with different output spaces. For example, six natural language tasks are considered in [3], including part-of-speech tagging, chunking, named entity recognition, semantic role labeling, language modeling, and semantically related word classification, and all those tasks take natural sentences as inputs.

11.3 Dual Learning, GANs and Autoencoder

As briefly described in Chap. 5.1.1 and shown in Fig. 11.1, GANs introduce a discriminator to help the training of the generator: the discriminator aims to differentiate generated samples from real samples, and the generator aims to fool the discriminator and maximize its error rate.

An autoencoder is a type of neural networks used for representation learning in an unsupervised manner. It aims to learn a representation (encoding) for a set of data, typically for dimensionality reduction, by training the network to ignore "noise" or irrelevant information in the data. As shown in Fig. 11.2, an autoencoder learns to copy its input to its output by using an encoder and a

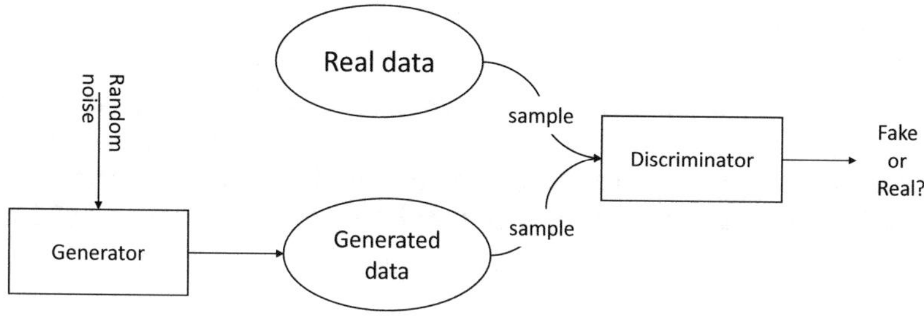

Fig. 11.1 Basic idea of GANs. Copied from Fig. 5.2

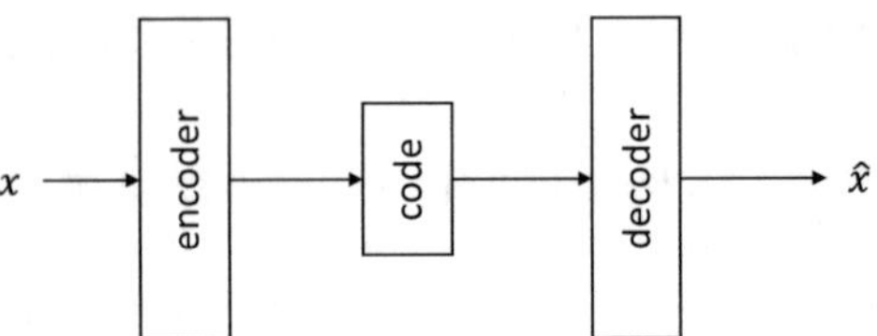

Fig. 11.2 Basic idea of autoencoders

decoder. The encoder maps the input into a hidden representation, and the decoder decodes the hidden representation and reconstructs the original input. Various techniques have been introduced to prevent autoencoders from learning the identity function, such as sparse autoencoders [14], denoising autoencoders [19], contractive autoencoders [16], and variational autoencoders [9].

In high level, dual learning looks similar to generative adversarial networks (GANs) and autoencoders, all of which involve joint training of two models. Dual learning is different from GANs and autoencoders in several aspects.

First, the joint training of two models in dual learning and autoencoders is in a cooperative manner, while that in GANs is in a competitive manner: the primal model and dual model help each other to achieve better accuracy, the encoder and decoder work together to minimize the reconstruction error, but the generator and discriminator in GANs have different objectives to optimize.

Second, while both dual learning and autoencoders share the reconstruction principle, more principles can be derived from structural duality in dual learning, such as the joint-probability principle (see Chaps. 7 and 8) and the marginal-probability principle (see Chap. 9).

Third, in dual learning, both the two models correspond to real world tasks, and after training both the improved models will be used. In GANs, the discriminator is introduced to help the training of the generator, and only the improved generator will be used in real world tasks. Similarly, in autoencoders after training only the encoder will be finally used for downstream tasks to output sparse or compact representations for data samples.

11.4 Dual Supervised Learning and Bayesian Ying-Yang Learning

A related concept to dual supervised learning is Bayesian Ying-Yang (BYY) learning,[1] or Ying-Yang machine [24, 25], which is a learning system with two pathways between the external observation domain X and its inner representation domain R. In BYY learning, the domain R and the pathway $R \to X$ is modeled by

[1]"Ying" is spelled "Yin" in Chinese Pin Yin. To make it harmony with Yang, Xu deliberately adopted the term "Ying-Yang" [22].

one subsystem system, which is called the Ying subsystem, while the domain X and the pathway $X \rightarrow R$ is modeled by another subsystem, which is called the Yang subsystem.

Both dual supervised learning and BYY learning take the forward and backward mappings into consideration. In dual learning, the two mappings are called primal and dual models, while in BYY learning they are called Ying and Yang subsystems. A slight difference is that in dual supervised learning, the primal model only considers the conditional distribution $p(Y|X)$ and the dual model only considers the conditional distribution $q(X|Y)$; in BYY learning, the Yang consists of the marginal distribution $p(X)$ modeling the probabilistic structure of the domain X and the conditional distribution $p(R|X)$ modeling the pathway $X \rightarrow R$, and similarly the Ying consists of the marginal distribution $q(R)$ for the domain R and the conditional distribution $q(X|R)$ for the pathway $R \rightarrow X$.

Dual supervised learning considers the hard constraint of the joint distribution on (X, Y):

$$p(X)p(Y|X) = q(Y)q(X|Y),$$

which is then convert to an objective to minimize in training:

$$\left[\log p(X) + \log p(Y|X) - \log q(Y) - \log q(X|Y)\right]^2.$$

In BYY learning, the Ying-Yang best harmony principle is to maximize the following harmony measure

$$H(p||q) = \int p(R|x)p(X)\ln[q(X|R)q(R)]dXdR. \tag{11.1}$$

Since BYY learning mainly considers continuous X and R, the above Ying-Yang best harmony principle does integration over X and R. Dual learning covers both discrete and continuous X and Y, and thus directly applies the constraint on each data sample without integration.

As BYY learning was proposed before the deep learning era [22], it is mainly studied for classical machine learning problems with shallow models, such as learning Gaussian mixture models [23] and (generalized) linear matrix systems [26]. Different from BYY learning, dual learning is mainly based deep neural networks and studied in complex real-world applications, such as machine translation, image translation, speech synthesis and recognition, question answering and generation, image classification and generation, etc.

11.5 Dual Reconstruction and Related Concepts

The dual reconstruction principle in dual learning, which aims to reconstruct an instance through the forward mapping and then the backward mapping, is related to several concepts studied in individual applications under different names.

The *forward-backward consistency* [7, 18] for point tracking shares the same idea as dual reconstruction in high level. Point tracking is a computer vision task that aims to estimate the location of a point in time $t + 1$ given its location in time t. In practice, tracking tends to fail when the points dramatically change appearance or disappear from the camera view. Kalal et al. [7] propose to leverage forward-backward consistency to detect tracking failures. The method works in three steps. (1) A tracker produces a forward trajectory by tracking the point forward in time. (2) A backward trajectory is obtained by backward tracking from the last frame to the first one. (3) If the two trajectories are significantly different, the forward trajectory is considered as a detection failure.

Zach et al. [29] utilize the observed redundancy in the hypothesized visual relations and chain reversible transformations over cycles in the graph induced by the relations to build suitable statistics for disambiguating visual relations in the graph. They call it *loop constraints.*

Cycle consistency has been studied in different problems, including shape matching [6], image co-segmentation [20], image alignment [31, 32], and image translation [33] as introduced in Chap. 5.3.2.

It is easy to see that the dual reconstruction principle and those consistency based concepts share the same idea in high level. In addition to this principle, structural duality implies other principles including the joint-probability principle and the marginal-probability principle, which have also been studied and explored in dual learning. We believe those principles should also be helpful for the above computer vision tasks.

References

1. Blum, A., & Mitchell, T. (1998). Combining labeled and unlabeled data with co-training. In *Proceedings of the Eleventh Annual Conference on Computational Learning Theory* (pp. 92–100).
2. Caruana, R. (1997). Multitask learning. *Machine Learning, 28*(1), 41–75.
3. Collobert, R., & Weston, J. (2008). A unified architecture for natural language processing: Deep neural networks with multitask learning. In *Proceedings of the 25th International Conference on Machine Learning* (pp. 160–167).
4. Evgeniou, T., & Pontil, M. (2004). Regularized multi-task learning. In *Proceedings of the Tenth ACM SIGKDD International Conference on Knowledge Discovery and Data Mining* (pp. 109–117).
5. He, D., Xia, Y., Qin, T., Wang, L., Yu, N., Liu, T.-Y., et al. (2016). Dual learning for machine translation. In *Advances in Neural Information Processing Systems* (pp. 820–828).
6. Huang, Q.-X., & Guibas, L. (2013). Consistent shape maps via semidefinite programming. In *Computer Graphics Forum* (vol. 32, pp. 177–186). Wiley Online Library.

7. Kalal, Z., Mikolajczyk, K., & Matas, J. (2010). Forward-backward error: Automatic detection of tracking failures. In *2010 20th International Conference on Pattern Recognition* (pp. 2756–2759). IEEE.
8. Kim, T., Cha, M., Kim, H., Lee, J. K., & Kim, J. (2017). Learning to discover cross-domain relations with generative adversarial networks. In *Proceedings of the 34th International Conference on Machine Learning-Volume 70* (pp. 1857–1865). JMLR.org.
9. Kingma, D. P., & Welling, M. (2013). Auto-encoding variational bayes. Preprint. arXiv:1312.6114.
10. Krogel, M.-A., & Scheffer, T. (2004). Multi-relational learning, text mining, and semi-supervised learning for functional genomics. *Machine Learning, 57*(1-2), 61–81.
11. Lin, J., Xia, Y., Qin, T., Chen, Z., & Liu, T.-Y. (2018). Conditional image-to-image translation. In *Proceedings of the IEEE Conference on Computer Vision and Pattern Recognition* (pp. 5524–5532).
12. Liu, P., Qiu, X., & Huang, X. (2016). Recurrent neural network for text classification with multi-task learning. In *Proceedings of the Twenty-Fifth International Joint Conference on Artificial Intelligence* (pp. 2873–2879).
13. Nigam, K., & Ghani, R. (2000). Analyzing the effectiveness and applicability of co-training. In *Proceedings of the Ninth International Conference on Information and Knowledge Management* (pp. 86–93).
14. Ranzato, M., Boureau, Y.-L., & Cun, Y. L. (2008). Sparse feature learning for deep belief networks. In *Advances in Neural Information Processing Systems* (pp. 1185–1192).
15. Ren, Y., Tan, X., Qin, T., Zhao, S., Zhao, Z., & Liu, T.-Y. (2019). Almost unsupervised text to speech and automatic speech recognition. In *International Conference on Machine Learning* (pp. 5410–5419).
16. Rifai, S., Vincent, P., Muller, X., Glorot, X., & Bengio, Y. (2011). Contractive auto-encoders: explicit invariance during feature extraction. In *Proceedings of the 28th International Conference on International Conference on Machine Learning* (pp. 833–840).
17. Ruder, S. (2017). An overview of multi-task learning in deep neural networks. Preprint. arXiv:1706.05098.
18. Sundaram, N., Brox, T., & Keutzer, K. (2010). Dense point trajectories by gpu-accelerated large displacement optical flow. In *European Conference on Computer Vision* (pp. 438–451). Springer.
19. Vincent, P., Larochelle, H., Lajoie, I., Bengio, Y., & Manzagol, P.-A. (2010). Stacked denoising autoencoders: Learning useful representations in a deep network with a local denoising criterion. *Journal of Machine Learning Research, 11*(Dec), 3371–3408.
20. Wang, F., Huang, Q., & Guibas, L. J. (2013). Image co-segmentation via consistent functional maps. In *Proceedings of the IEEE International Conference on Computer Vision* (pp. 849–856).
21. Wang, Y., Xia, Y., He, T., Tian, F., Qin, T., Zhai, C. X., et al. (2019). Multi-agent dual learning. In *7th International Conference on Learning Representations, ICLR 2019*.
22. Xu, L. (1995). Bayesian-kullback coupled ying-yang machines: Unified learnings and new results on vector quantization. In *Proceedings of ICONIP95, Oct 30-Nov 3, 1995, Beijing, China, 1995* (pp. 977–988).
23. Xu, L. (1998). Rbf nets, mixture experts, and bayesian ying–yang learning. *Neurocomputing, 19*(1-3), 223–257.
24. Xu, L. (2004). Bayesian ying yang learning (i): a unified perspective for statistical modeling. In *Intelligent Technologies for Information Analysis* (pp. 615–659). Springer.
25. Xu, L. (2004). Bayesian ying yang learning (ii): A new mechanism for model selection and regularization. In *Intelligent Technologies for Information Analysis* (pp. 661–706). Springer.
26. Xu, L. (2015). Further advances on bayesian ying-yang harmony learning. In *Applied Informatics* (vol. 2, p. 5). Springer.

27. Xu, J., Tan, X., Ren, Y., Qin, T., Li, J., Zhao, S., et al. (2020). Lrspeech: Extremely low-resource speech synthesis and recognition. In *Proceedings of the 26th Acm Sigkdd International Conference on Knowledge Discovery and Data Mining*.
28. Yi, Z., Zhang, H., Tan, P., & Gong, M. (2017). Dualgan: Unsupervised dual learning for image-to-image translation. In *Proceedings of the IEEE International Conference on Computer Vision* (pp. 2849–2857).
29. Zach, C., Klopschitz, M., & Pollefeys, M. (2010). Disambiguating visual relations using loop constraints. In *2010 IEEE Computer Society Conference on Computer Vision and Pattern Recognition* (pp. 1426–1433). IEEE.
30. Zhang, Z., Luo, P., Loy, C. C., & Tang, X. (2014). Facial landmark detection by deep multi-task learning. In *European Conference on Computer Vision* (pp. 94–108). Springer.
31. Zhou, T., Jae Lee, Y., Yu, S. X., & Efros, A. A. (2015). Flowweb: Joint image set alignment by weaving consistent, pixel-wise correspondences. In *Proceedings of the IEEE Conference on Computer Vision and Pattern Recognition* (pp. 1191–1200).
32. Zhou, T., Krahenbuhl, P., Aubry, M., Huang, Q., & Efros, A. A. (2016). Learning dense correspondence via 3d-guided cycle consistency. In *Proceedings of the IEEE Conference on Computer Vision and Pattern Recognition* (pp. 117–126).
33. Zhu, J.-Y., Park, T., Isola, P., & Efros, A. A. (2017). Unpaired image-to-image translation using cycle-consistent adversarial networks. In *Proceedings of the IEEE International Conference on Computer Vision* (pp. 2223–2232).

Part V
Summary and Outlook

In the last part of this book, we provide a brief summary to current research on dual learning and outlook future research directions.

Chapter 12
Summary and Outlook

This book is just a stage-wise summary of the hot research field of dual learning. Given the fast development of the field, we can foresee that many new algorithms and theories will be developed. We hope that this book will motivate more people to work on dual learning so as to generate more impact in both machine learning and AI communities. Furthermore, we hope that this book will help industry practitioners to enrich their weapons so as to better solve their problems.

12.1 Summary

In this book, we have given an overview to dual learning, including basic principles, different learning settings, and diverse applications.

There are two main principles to leverage structural duality between machine learning tasks.

- The dual reconstruction principle (Chap. 4.2), which requires that a sample x should be reconstructed if it is processed by the primal/dual model first and then the dual/primal model. That is,

$$x = g(f(x)), \quad y = f(g(y)).$$

  Semi-supervised and unsupervised dual learning algorithms are mainly based on this principle.
- The probability principle, which builds probability connection between the primal and dual models. There are two sub principles. The first one is the joint-probability principle (Chap. 7.1), stating that the joint probability of a (x, y) pair computed using the primal model should equal to that using the dual model:

$$P(x, y) = P(x)P(y|x; f) = P(y)P(x|y; g).$$

T. Qin, *Dual Learning*, https://doi.org/10.1007/978-981-15-8884-6_12

The second one is the marginal-probability principle (Sect. 9.1), stating that the marginal-probability of a sample y can be computed using both models:

$$P(y) = \mathbb{E}_{x \sim P(x)} P(y|x; f) = \mathbb{E}_{x \sim P(x|y;g)} P(y|x; f) \frac{P(x)}{P(x|y; g)}.$$

The joint-probability principle is mainly used in supervised learning with labeled data and in inference, and the marginal-probability principle is usually used to learning from unlabeled data.

We have introduced dual learning in four major learning settings:

- Dual semi-supervised learning (Sects. 4.3, 4.5, Chaps. 6, 9) conducts dual learning using both labeled and unlabeled data.
- Dual unsupervised learning (Sect. 4.4, Chap. 5) uses only unlabeled data for training.
- Dual supervised learning (Chap. 7) uses only labeled data for training.
- Dual inference (Chap. 8) leverages structure duality for inference.

Dual learning has been studied in and applied to many applications, spanning from machine translation, image translation, speech synthesis and generation, to question answering and generation, code summarization and generation, image classification and generation, text summarization, sentiment analysis, etc.

To give a high-level understanding on the dual learning research, we categorize representative works based on their principles, learning settings and targeted applications in Fig. 12.1.

12.2 Future Directions

Although dual learning has been studied in multiple learning settings and applied to many applications, there are still many open problems worthy of explorations.

12.2.1 More Learning Settings and Applications

The duality between machine learning tasks has been studied in several popular machine learning settings, including unsupervised setting, semi-supervised setting, supervised setting, and inference setting. We believe it has potential to boost other machine learning settings. Here we give a few examples.

- Recall that the basic motivation of dual learning when it was formally proposed [15] is to address the shortage of labeled data for machine learning algorithms, especially deep neural networks, and to better utilize unlabeled data. There are some other well-known approaches to learn from unlabeled

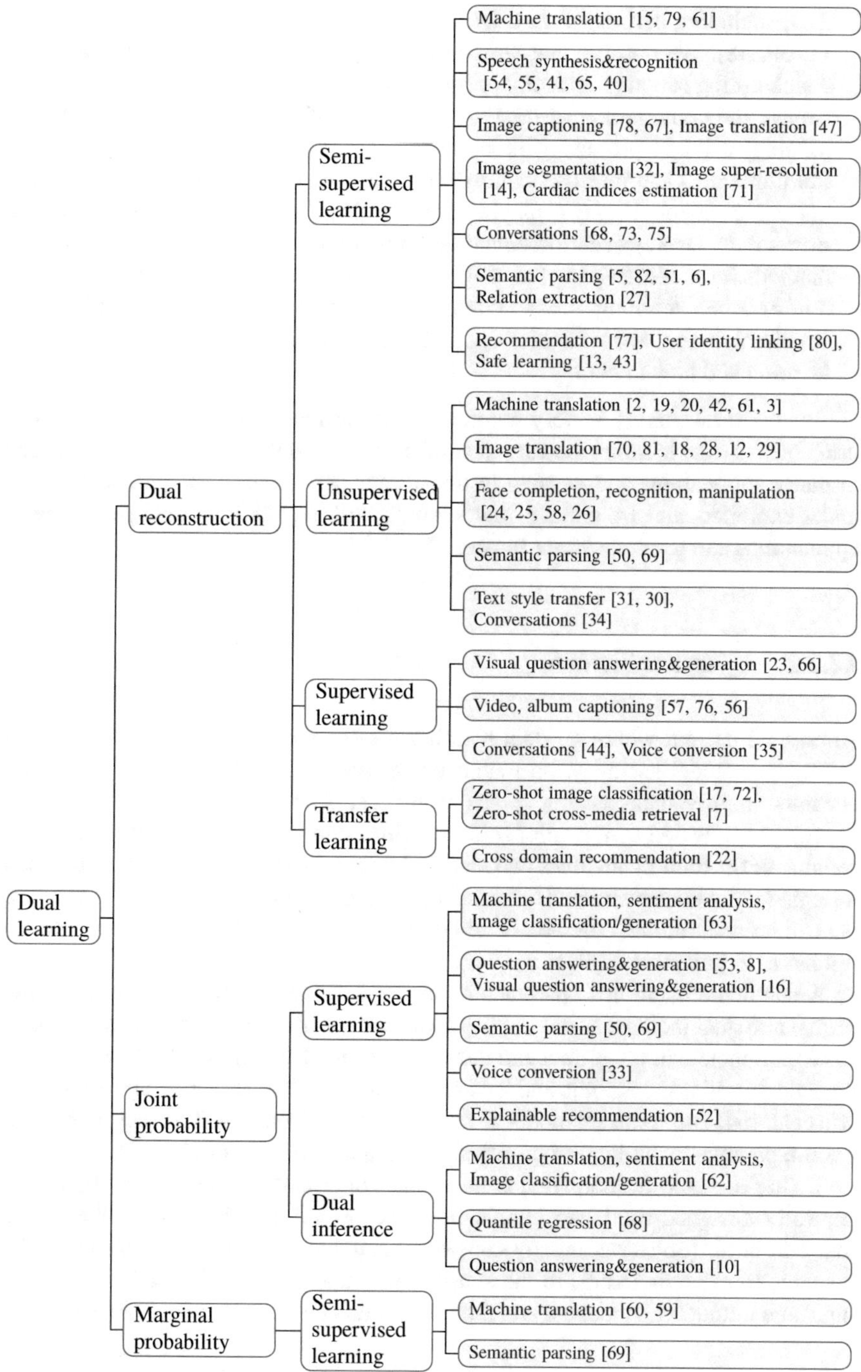

Fig. 12.1 Taxonomy of dual learning works

data, including self-supervised learning [37, 38, 49, 73] and pre-training[1] [9, 11, 39, 48]. We observe that structural duality is not used in those works and dual learning is complementary to self-supervised learning and pre-training. We believe the combination of dual learning and self-supervised learning (or pre-training) will be a promising direction.

- Reinforcement learning, especially deep reinforcement learning, has made amazing achievements in game playing [25, 45, 46]. While multiple dual learning works have leverage reinforcement learning algorithms such as policy gradient methods for model training [15, 40], dual learning has not been explored to boost reinforcement learning. It is an interesting direction to investigate what kinds of structural duality and what kinds of principles can be used to help reinforcement learning and how to help.

As structural duality is very common in many real world tasks and the basic principles of dual learning are very general, although dual learning has been studied in many applications and verified to boost their performance, its potential is still under exploited, and we believe dual learning will play an important role in more applications and generate larger impact.

12.2.2 *Efficient Training*

Almost all current works on dual learning focus on leveraging structural duality to improve model accuracy, no matter using labeled data or unlabeled data. The accuracy improvement usually comes at the cost of increased training complexity in terms of time and computation, either because more (unlabeled) data is used for training or the training involves two models (the primal and dual models) rather than a single model in conventional machine learning algorithms. A challenge is whether we can reduce training complexity while improving model accuracy, or at least do not increase training complexity.

A candidate solution is parameter sharing, i.e., sharing the parameters in the primal and dual models, as did in [64]. Intuitively, with parameter sharing, we have fewer parameters in the primal and dual models, and it is very likely the training of the models will take less time and fewer data samples to convergence. Unfortunately, Xia et al. [64] still focus on model accuracy instead of training efficiency, but we do believe parameter sharing is a promising approach to improve training efficiency.

Taking one step further, another candidate solution could be using a single and carefully designed model to handle both the primal and dual tasks. We notice that Niu et al. [36] have already conducted some studies along this direction for machine translation, thanks to the symmetry of machine translation (i.e., both the input and output are sequences and the encoder and decoder share the same network

[1] Pre-training can also be viewed as a special kind of self-supervised learning.

structure). For general machine learning tasks (e.g., image classification), the input and output are usually of different formats. We need to design or consider new network architectures. For example, a special kind of neural networks, invertible neural networks (INNs) [1, 4], is well suited for this purpose. Unlike classical neural networks, which only focus on the forward mapping from $x \in \mathcal{X}$ to $y \in \mathcal{Y}$ and all the information irrelevant to y is lost in training, INNs focus on learning the forward process while using additional latent output variables to capture the information otherwise lost. Due to invertibility, a model of the corresponding inverse process is learned implicitly. We expect better training efficiency can be achieved by leveraging the idea of invertible neural networks.

12.2.3 Theoretical Understanding

Many dual learning algorithms have been designed for different settings and diverse applications, but our theoretical understanding on dual learning is still very limited. According to our knowledge, only few research efforts have been put on this direction, such as [12, 62, 63, 80], and the results from those studies are still far from a good understanding of dual learning.

- The theoretical studies in [63] and [62] are just direct and simple application of previous theoretical results on conventional machine learning algorithms and do not bring much insight to dual learning.
- The analysis in [12] on dual unsupervised learning for image to image translation heavily relies on the simplicity hypothesis, which unfortunately is not theoretically proved and even worse does not hold in some real world applications, as mentioned in Sect. 10.3.
- While the analysis in [80] on dual semi-supervised learning for machine translation does not make strong assumptions, the results (e.g., Theorem 10.12) are very complex and it is not clear whether this kind of analysis can be easily extended to other applications and learning settings.

Therefore, we still need significant theoretical efforts to deeply and better understand dual learning. In principle, we would like to weaken our assumptions and get general results/insight that work for different learning settings and diverse applications.

References

1. Ardizzone, L., Kruse, J., Rother, C., & Köthe, U. (2018). Analyzing inverse problems with invertible neural networks. In *International Conference on Learning Representations*.
2. Artetxe, M., Labaka, G., Agirre, E., & Cho, K. (2018). Unsupervised neural machine translation. In *6th International Conference on Learning Representations*.

3. Bai, X., Zhang, Y., Cao, H., & Zhao, T. (2019). Duality regularization for unsupervised bilingual lexicon induction. Preprint. arXiv:1909.01013.
4. Behrmann, J., Grathwohl, W., Chen, R. T. Q., Duvenaud, D., & Jacobsen, J.-H. (2019). Invertible residual networks. In *International Conference on Machine Learning* (pp. 573–582).
5. Cao, R., Zhu, S., Liu, C., Li, J., & Yu, K. (2019). Semantic parsing with dual learning. In *Proceedings of the 57th Annual Meeting of the Association for Computational Linguistics* (pp. 51–64).
6. Cao, R., Zhu, S., Yang, C., Liu, C., Ma, R., Zhao, Y., et al. (2020). Unsupervised dual paraphrasing for two-stage semantic parsing. Preprint. arXiv:2005.13485.
7. Chi, J., & Peng, Y. (2019). Zero-shot cross-media embedding learning with dual adversarial distribution network. *IEEE Transactions on Circuits and Systems for Video Technology, 30*(4), 1173–1187.
8. Cui, S., Lian, R., Jiang, D., Song, Y., Bao, S., & Jiang, Y. (2019). Dal: Dual adversarial learning for dialogue generation. In *Proceedings of the Workshop on Methods for Optimizing and Evaluating Neural Language Generation* (pp. 11–20).
9. Devlin, J., Chang, M.-W., Lee, K., & Toutanova, K. (2019). Bert: Pre-training of deep bidirectional transformers for language understanding. In *NAACL-HLT (1)*.
10. Duan, N., Tang, D., Chen, P., & Zhou, M. (2017). Question generation for question answering. In *Proceedings of the 2017 Conference on Empirical Methods in Natural Language Processing* (pp. 866–874).
11. Erhan, D., Courville, A., Bengio, Y., & Vincent, P. (2010). Why does unsupervised pre-training help deep learning? In *Proceedings of the Thirteenth International Conference on Artificial Intelligence and Statistics* (pp. 201–208).
12. Galanti, T., Wolf, L., & Benaim, S. (2018). The role of minimal complexity functions in unsupervised learning of semantic mappings. In *ICLR 2018: International Conference on Learning Representations 2018*.
13. Gan, H., Li, Z., Fan, Y. & Luo, Z. (2017). Dual learning-based safe semi-supervised learning. *IEEE Access, 6*, 2615–2621.
14. Guo, Y., Chen, J., Wang, J., Chen, Q., Cao, J., Deng, Z., et al. (2020). Closed-loop matters: Dual regression networks for single image super-resolution. In *Proceedings of the IEEE/CVF Conference on Computer Vision and Pattern Recognition* (pp. 5407–5416).
15. He, D., Xia, Y., Qin, T., Wang, L., Yu, N., Liu, T.-Y., et al. (2016). Dual learning for machine translation. In *Advances in Neural Information Processing Systems* (pp. 820–828).
16. He, S., Han, C., Han, G., & Qin, J. (2019). Exploring duality in visual question-driven top-down saliency. *IEEE Transactions on Neural Networks and Learning Systems.*
17. Huang, H., Wang, C., Yu, P. S., & Wang, C.-D. (2019). Generative dual adversarial network for generalized zero-shot learning. In *Proceedings of the IEEE conference on Computer Vision and Pattern Recognition* (pp. 801–810).
18. Kim, T., Cha, M., Kim, H., Lee, J. K., & Kim, J. (2017). Learning to discover cross-domain relations with generative adversarial networks. In *Proceedings of the 34th International Conference on Machine Learning-Volume 70* (pp. 1857–1865). JMLR.org.
19. Lample, G., Conneau, A., Denoyer, L., & Ranzato, M. (2018). Unsupervised machine translation using monolingual corpora only. In *6th International Conference on Learning Representations, ICLR 2018*.
20. Lample, G., Ott, M., Conneau, A., Denoyer, L., & Ranzato, M. (2018). Phrase-based & neural unsupervised machine translation. In *Proceedings of the 2018 Conference on Empirical Methods in Natural Language Processing, Brussels, Belgium, October 31 - November 4, 2018* (pp. 5039–5049).
21. Li, P., & Tuzhilin, A. (2020). Ddtcdr: Deep dual transfer cross domain recommendation. In *Proceedings of the 13th International Conference on Web Search and Data Mining* (pp. 331–339).
22. Li, Z., Hu, Y., & He, R. (2017). Learning disentangling and fusing networks for face completion under structured occlusions. Preprint. arXiv:1712.04646.

23. Li, Y., Duan, N., Zhou, B., Chu, X., Ouyang, W., Wang, X., et al. (2018). Visual question generation as dual task of visual question answering. In *Proceedings of the IEEE Conference on Computer Vision and Pattern Recognition* (pp. 6116–6124).
24. Li, Z., Hu, Y., Zhang, M., Xu, M., & He, R. (2018). Protecting your faces: Meshfaces generation and removal via high-order relation-preserving cyclegan. In *2018 International Conference on Biometrics (ICB)* (pp. 61–68). IEEE.
25. Li, J., Koyamada, S., Ye, Q., Liu, G., Wang, C., Yang, R., et al. (2020). Suphx: Mastering mahjong with deep reinforcement learning. Preprint. arXiv:2003.13590.
26. Li, Z., Hu, Y., He, R., & Sun, Z. (2020). Learning disentangling and fusing networks for face completion under structured occlusions. *Pattern Recognition, 99*, 107073.
27. Lin, J., Xia, Y., Qin, T., Chen, Z., & Liu, T.-Y. (2018). Conditional image-to-image translation. In *Proceedings of the IEEE Conference on Computer Vision and Pattern Recognition* (pp. 5524–5532).
28. Lin, H., Yan, J., Qu, M., & Ren, X. (2019). Learning dual retrieval module for semi-supervised relation extraction. In *The World Wide Web Conference* (pp. 1073–1083).
29. Lin, J., Xia, Y., Wang, Y., Qin, T., & Chen, Z. (2019). Image-to-image translation with multi-path consistency regularization. In *Proceedings of the Twenty-Eighth International Joint Conference on Artificial Intelligence* (pp. 2980–2986).
30. Luo, P., Wang, G., Lin, L., & Wang, X. (2017). Deep dual learning for semantic image segmentation. In *Proceedings of the IEEE International Conference on Computer Vision* (pp. 2718–2726).
31. Luo, F., Li, P., Yang, P., Zhou, J., Tan, Y., Chang, B., et al. (2019) Towards fine-grained text sentiment transfer. In *Proceedings of the 57th Annual Meeting of the Association for Computational Linguistics* (pp. 2013–2022).
32. Luo, F., Li, P., Zhou, J., Yang, P., Chang, B., Sun, X., et al. (2019). A dual reinforcement learning framework for unsupervised text style transfer. In *Proceedings of the 28th International Joint Conference on Artificial Intelligence* (pp. 5116–5122). AAAI Press.
33. Luo, Z., Chen, J., Takiguchi, T., & Ariki, Y. (2019). Emotional voice conversion using dual supervised adversarial networks with continuous wavelet transform f0 features. *IEEE/ACM Transactions on Audio, Speech, and Language Processing, 27*(10), 1535–1548.
34. Meng, C., Ren, P., Chen, Z., Sun, W., Ren, Z., Tu, Z., et al. (2020). Dukenet: A dual knowledge interaction network for knowledge-grounded conversation. In *Proceedings of the 43rd International ACM SIGIR Conference on Research and Development in Information Retrieval* (pp. 1151–1160).
35. Miyoshi, H., Saito, Y., Takamichi, S., & Saruwatari, H. (2017). Voice conversion using sequence-to-sequence learning of context posterior probabilities. In *Proc. Interspeech 2017* (pp. 1268–1272).
36. Niu, X., Denkowski, M., & Carpuat, M. (2018). Bi-directional neural machine translation with synthetic parallel data. In *Proceedings of the 2nd Workshop on Neural Machine Translation and Generation* (pp. 84–91).
37. Pathak, D., Krahenbuhl, P., Donahue, J., Darrell, T., & Efros, A. A. (2016). Context encoders: Feature learning by inpainting. In *Proceedings of the IEEE Conference on Computer Vision and Pattern Recognition* (pp. 2536–2544).
38. Pathak, D., Agrawal, P., Efros, A. A., & Darrell, T. (2017). Curiosity-driven exploration by self-supervised prediction. In *Proceedings of the IEEE Conference on Computer Vision and Pattern Recognition Workshops* (pp. 16–17).
39. Radford, A., Narasimhan, K., Salimans, T., & Sutskever, I. (2018). Improving language understanding by generative pre-training.
40. Radzikowski, K., Nowak, R., Wang, L., & Yoshie, O. (2019). Dual supervised learning for non-native speech recognition. *EURASIP Journal on Audio, Speech, and Music Processing, 2019*(1), 3.
41. Ren, Y., Tan, X., Qin, T., Zhao, S., Zhao, Z., & Liu, T.-Y. (2019). Almost unsupervised text to speech and automatic speech recognition. In *International Conference on Machine Learning* (pp. 5410–5419).

42. Sestorain, L., Ciaramita, M., Buck, C., & Hofmann, T. (2018). Zero-shot dual machine translation. Preprint. arXiv:1805.10338.
43. She, Q., Zou, J., Luo, Z., Nguyen, T., Li, R., & Zhang, Y. (2020). Multi-class motor imagery eeg classification using collaborative representation-based semi-supervised extreme learning machine. *Medical & Biological Engineering & Computing*, 1–12.
44. Shen, L., & Feng, Y. (2020). Cdl: Curriculum dual learning for emotion-controllable response generation. Preprint. arXiv:2005.00329.
45. Silver, D., Huang, A., Maddison, C. J., Guez, A., Sifre, L., Van Den Driessche, G., et al. (2016). Mastering the game of go with deep neural networks and tree search. *Nature, 529*(7587), 484.
46. Silver, D., Hubert, T., Schrittwieser, J., Antonoglou, I., Lai, M., Guez, A., et al. (2018). A general reinforcement learning algorithm that masters chess, shogi, and go through self-play. *Science, 362*(6419), 1140–1144.
47. Song, J., Pang, K., Song, Y.-Z., Xiang, T., & Hospedales, T. M. (2018). Learning to sketch with shortcut cycle consistency. In *Proceedings of the IEEE Conference on Computer Vision and Pattern Recognition* (pp. 801–810).
48. Song, K., Tan, X., Qin, T., Lu, J., & Liu, T.-Y. (2019). Mass: Masked sequence to sequence pre-training for language generation. In *International Conference on Machine Learning* (pp. 5926–5936).
49. Srivastava, N., Mansimov, E., & Salakhudinov, R. (2015). Unsupervised learning of video representations using lstms. In *International Conference on Machine Learning* (pp. 843–852).
50. Su, S.-Y., Huang, C.-W., & Chen, Y.-N. (2019). Dual supervised learning for natural language understanding and generation. In *Proceedings of the 57th Annual Meeting of the Association for Computational Linguistics* (pp. 5472–5477).
51. Su, S.-Y., Huang, C.-W., & Chen, Y.-N. (2020). Towards unsupervised language understanding and generation by joint dual learning. In *ACL 2020: 58th Annual Meeting of the Association for Computational Linguistics* (pp. 671–680).
52. Sun, Y., Tang, D., Duan, N., Qin, T., Liu, S., Yan, Z., et al. (2019). Joint learning of question answering and question generation. *IEEE Transactions on Knowledge and Data Engineering*.
53. Sun, P., Wu, L., Zhang, K., Fu, Y., Hong, R., & Wang, M. (2020). Dual learning for explainable recommendation: Towards unifying user preference prediction and review generation. In *Proceedings of The Web Conference 2020* (pp. 837–847).
54. Tjandra, A., Sakti, S., & Nakamura, S. (2017). Listening while speaking: Speech chain by deep learning. In *Automatic Speech Recognition and Understanding Workshop (ASRU), 2017 IEEE* (pp. 301–308). IEEE.
55. Tjandra, A., Sakti, S., & Nakamura, S. (2018). Machine speech chain with one-shot speaker adaptation. In *Proc. Interspeech 2018* (pp. 887–891).
56. Wang, S., & Peng, G. (2019). Weakly supervised dual learning for facial action unit recognition. *IEEE Transactions on Multimedia, 21*(12), 3218–3230.
57. Wang, B., Ma, L., Zhang, W., & Liu, W. (2018). Reconstruction network for video captioning. In *Proceedings of the IEEE Conference on Computer Vision and Pattern Recognition* (pp. 7622–7631).
58. Wang, Y., Xia, Y., Zhao, L., Bian, J., Qin, T., Liu, G., et al. (2018). Dual transfer learning for neural machine translation with marginal distribution regularization. In *Thirty-Second AAAI Conference on Artificial Intelligence*.
59. Wang, B., Ma, L., Zhang, W., Jiang, W., & Zhang, F. (2019). Hierarchical photo-scene encoder for album storytelling. In *Proceedings of the AAAI Conference on Artificial Intelligence* (vol. 33, pp. 8909–8916).
60. Wang, Y., Xia, Y., Zhao, L., Bian, J., Qin, T., Chen, E., et al. (2019). Semi-supervised neural machine translation via marginal distribution estimation. *IEEE/ACM Transactions on Audio, Speech, and Language Processing, 27*(10), 1564–1576.
61. Wang, Y., Xia, Y., He, T., Tian, F., Qin, T., Zhai, C. X., et al. (2019). Multi-agent dual learning. In *7th International Conference on Learning Representations, ICLR 2019*.

62. Xia, Y., Bian, J., Qin, T., Yu, N., & Liu, T.-Y. (2017). Dual inference for machine learning. In *Proceedings of the 26th International Joint Conference on Artificial Intelligence* (pp. 3112–3118).
63. Xia, Y., Qin, T., Chen, W., Bian, J., Yu, N., & Liu, T.-Y. (2017). Dual supervised learning. In *Proceedings of the 34th International Conference on Machine Learning-Volume 70* (pp. 3789–3798). JMLR.org.
64. Xia, Y., Tan, X., Tian, F., Qin, T., Yu, N., & Liu, T.-Y. (2018). Model-level dual learning. In *International Conference on Machine Learning* (pp. 5383–5392).
65. Xu, X., Song, J., Lu, H., He, L., Yang, Y., & Shen, F. (2018). Dual learning for visual question generation. *2018 IEEE International Conference on Multimedia and Expo (ICME)* (pp. 1–6).
66. Xu, J., Tan, X., Ren, Y., Qin, T., Li, J., Zhao, S., et al. (2020). Lrspeech: Extremely low-resource speech synthesis and recognition. In *Proceedings of the 26th acm Sigkdd International Conference on Knowledge Discovery and Data Mining*.
67. Yang, M., Zhao, Z., Zhao, W., Chen, X., Zhu, J., Zhou, L., et al. (2017). Personalized response generation via domain adaptation. In *Proceedings of the 40th International ACM SIGIR Conference on Research and Development in Information Retrieval* (pp. 1021–1024).
68. Yang, M., Zhao, W., Xu, W., Feng, Y., Zhao, Z., Chen, X., et al. (2018). Multitask learning for cross-domain image captioning. *IEEE Transactions on Multimedia, 21*(4), 1047–1061.
69. Ye, H., Li, W., & Wang, L. (2019) Jointly learning semantic parser and natural language generator via dual information maximization. In *Proceedings of the 57th Annual Meeting of the Association for Computational Linguistics* (pp. 2090–2101).
70. Yi, Z., Zhang, H., Tan, P., & Gong, M. (2017). Dualgan: Unsupervised dual learning for image-to-image translation. In *Proceedings of the IEEE International Conference on Computer Vision* (pp. 2849–2857).
71. Yu, C., Gao, Z., Zhang, W., Yang, G., Zhao, S., Zhang, H., et al. (2020). Multitask learning for estimating multitype cardiac indices in mri and ct based on adversarial reverse mapping. *IEEE Transactions on Neural Networks and Learning Systems*, 1–14. https://doi.org/10.1109/TNNLS.2020.2984955.
72. Zhang, Z., & Yang, J. (2018). Dual learning based multi-objective pairwise ranking. In *2018 International Joint Conference on Neural Networks (IJCNN)* (pp. 1–7). IEEE.
73. Zhang, R., Isola, P., & Efros, A. A. (2016). Colorful image colorization. In *European Conference on Computer Vision* (pp. 649–666). Springer.
74. Zhang, H., Lan, Y., Guo, J., Xu, J., & Cheng, X. (2018). Reinforcing coherence for sequence to sequence model in dialogue generation. In *Proceedings of the 27th International Joint Conference on Artificial Intelligence* (pp. 4567–4573).
75. Zhang, C., Lyu, X., & Tang, Z. (2019). Tgg: Transferable graph generation for zero-shot and few-shot learning. In *Proceedings of the 27th ACM International Conference on Multimedia* (pp. 1641–1649).
76. Zhang, S., & Bansal, M. (2019). Addressing semantic drift in question generation for semi-supervised question answering. In *Proceedings of the 2019 Conference on Empirical Methods in Natural Language Processing and the 9th International Joint Conference on Natural Language Processing (EMNLP-IJCNLP)* (pp. 2495–2509).
77. Zhang, W., Wang, B., Ma, L., & Liu, W. (2019). Reconstruct and represent video contents for captioning via reinforcement learning. *IEEE Transactions on Pattern Analysis and Machine Intelligence*, 1. https://doi.org/10.1109/TPAMI.2019.2920899
78. Zhao, W., Xu, W., Yang, M., Ye, J., Zhao, Z., Feng, Y., et al. (2017). Dual learning for cross-domain image captioning. In *Proceedings of the 2017 ACM on Conference on Information and Knowledge Management* (pp. 29–38).
79. Zhou, F., Liu, L., Zhang, K., Trajcevski, G., Wu, J., & Zhong, T. (2018). Deeplink: A deep learning approach for user identity linkage. In *IEEE INFOCOM 2018-IEEE Conference on Computer Communications* (pp. 1313–1321). IEEE.

80. Zhao, Z., Xia, Y., Qin, T., Xia, L., & Liu, T.-Y. (2020). Dual learning: Theoretical study and an algorithmic extension. Preprint. arXiv:2005.08238.
81. Zhu, J.-Y., Park, T., Isola, P., & Efros, A. A. (2017). Unpaired image-to-image translation using cycle-consistent adversarial networks. In *Proceedings of the IEEE International Conference on Computer Vision* (pp. 2223–2232).
82. Zhu, S., Cao, R., & Yu, K. (2020). Dual learning for semi-supervised natural language understanding. *IEEE Transactions on Audio, Speech, and Language Processing*.

Printed in the United States
by Baker & Taylor Publisher Services